Pendulums

Jared O'Keefe

Dowsing with pendulums and divining-rods affords movement between the physical and spiritual realms. The pendulum is used for getting direction in practical matters (the whereabouts of water, oil, etc.), and abstract issues (Earth energies and future prediction). The divining-rod reacts to the vibrations around us, providing answers and spiritual direction. This book teaches us how to apply these methods for our benefit and for entering a new spiritual age.

Jared O'Keefe's interest in dowsing began when, as a child, he watched a water- diviner at work. O'Keefe uses pendulums and divining rods to predict the future professionally.

ASTROLOG COMPLETE GUIDES SERIES

The Complete Guide to Coffee Grounds and Tea Leaf Reading
Sara Zed

The Complete Guide to Palmistry
Batia Shorek

The Complete Guide to Tarot Reading
Hali Morag

Crystals - Types, Use and Meaning
Connie Islin

The Dictionary of Dreams
Eili Goldberg

Meditation: The Journey to Your Inner World
Eidan Or

Playing Cards: Predicting Your Future
Hali Morag

Day-by-Day Numerology
Lia Robin

Using Astrology To Choose Your Partner
Amanda Starr

The I Ching
Nizan Weisman

Pendulums
Jared O'Keefe

Channeling
Roxanne McGuire

What Moles Tell You About Yourself
Pietro Santini

Secrets of the Body: Your Character and Future Revealed
Jocelyne Cooke

Secrets of the Face: Your Character and Future Revealed
Jocelyne Cooke

Astrology
Amanda Starr

Day-by-Day Wicca
Tabatha Jennings

Pendulums

Jared O'Keefe

Astrolog Publishing House

P.O. Box 1123, Hod Hasharon 45111, Israel

Tel: 972-9-7412044

Fax: 972-9-7442714

E-Mail: info@astrolog.co.il

Astrolog Web Site: www.astrolog.co.il

© Jared O'Keefe 1999

ISBN 965-494-090-6

All rights reserved. No part of this publication may be reproduced, stored in a retrieval system, or transmitted, in any form or by any means, electronic, mechanical, photocopying, recording or otherwise, without the prior permission of the publisher.

Published by Astrolog Publishing House 2000

Printed in Israel

10 9 8 7 6 5 4 3 2 1

CONTENTS

1. What is a pendulum? 7
2. How do the pendulum and the divining rod work? 10
3. How to use the pendulum: Methods for tuning and use of the pendulum 16
4. Rules for using the pendulum: Tuning into an objective - How to start 28
5. Preparation of the pendulum 32
6. The pendulum in romance 38
7. The pendulum in work and career 45
8. Using the pendulum for diagnosis and healing 50
9. The pendulum and our spiritual development 64
10. The pendulum and astrology 69
11. Using the pendulum for healing 73
12. Using the pendulum throughout the house 93
13. Advanced uses of the pendulum 106
14. The dowsing rod 116

This publication contains the opinions and ideas of its author. It is intended to provide helpful and informative material on the subject matter covered. It is sold with the understanding that the authors and publisher are not engaged in rendering professional services in the book. If the reader requires personal assistance or advice, a competent professional should be consulted.

The author, book producer, and publisher specifically disclaim any responsibility for any liability, loss or risk, personal or otherwise, which is incurred as a consequence, directly or indirectly, of the use and application of any of the contents of this book.

1

What is a pendulum?

The pendulum is a small weighted object suspended from a string that is held between the thumb and index finger. The movement of the object indicates positive or negative energy.

This simple-looking item has amazing and unusual capabilities. It can measure our internal energies, as well as our external energy fields, and can be used for many different purposes. The pendulum's sensitivity and accuracy make it a multi-functional tool in all fields of life, from discovering mineral deposits to making medical diagnoses.

Use of the pendulum is called *radiesthaesia* (sensitivity to radiation), and has been known and widespread throughout the world for hundreds of years; in fact, it was already known to some people thousands of years ago. In this book, a great variety of methods and uses of the pendulum will be described, encompassing all areas of life. Using the pendulum is very simple, and with a little practice and effort, amazing and far-reaching results can be obtained.

The history of the pendulum

When the use of the pendulum began is not known for certain. It is known to be a refinement of the divining rod, also called the divine rod, pictured in primitive paintings from 6000 BCE that were discovered in a cave in the Sahara Desert. The Egyptians and Chinese were familiar with the methods of using a divining rod, as were the Greeks and Romans. It is unclear when exactly the use of the divining rod was refined into the pendulum, which works on similar principles, but it is known that at the beginning of the century, its use became common and familiar. Bouly, a French priest from the village Haderlot, coined the term *radiesthaesia* to describe the use of the pendulum. This term is composed of the Latin word *radius*, meaning "radiation," and the Greek word *esthaesis*, meaning "sensitivity."

Following World War I, the use of the pendulum spread greatly after Bouly and another priest, Mermet, organized a series of congresses to increase knowledge about and scientific research on the many uses of the pendulum, with great emphasis placed on its medical uses. Mermet, who devoted a great deal of time to researching the pendulum and used it to help many people, is known throughout the world as the most famous pendulum expert that ever lived. His fields of activity encompassed many different areas, from finding lost children to helping the Vatican solve archaeological enigmas.

In the United States, use of the pendulum blossomed following the publication of a book by the English physician, Dr. Albert Abrams, in 1922, in which he described the use of the pendulum primarily for medical, therapeutic, and diagnostic purposes.

Andre Bovis of France discovered additional powers of the pendulum when he invented techniques for using the pendulum to gauge the freshness and quality of foods.

In 1947, Bruce Copen, an English manufacturer and marketer of pendulums, came out with an amazing statement: He claimed that ninety percent of the people in the world have the ability to utilize the pendulum in everyday life. He predicted that if half that number of people in fact exploited that ability and practiced using the pendulum, they could even use it for purposes of science and research.

The most famous and well-known researcher to go into the use and possibilities of the pendulum in depth was the Nobel Prize winner, Dr. Alexis Carrel. Over 50 years ago, while doing research at the reknowned Rockefeller Institute in New York, Dr. Carrel discovered the considerable ability of radiesthaesia to change the entire world. He expressed his scientific opinion in an unequivocal manner. The physiologist must investigate and discover the personal characteristics of each patient, as well as his contribution to the course of the illness, his sensitivity to pain, the general state of all his physical functions, his past, and his future. The physiologist must maintain open-mindedness, and freedom from any preconceptions that are the result of outmoded research methods. Therefore he must bear in mind that radiesthaesia deserves serious consideration.

And indeed, one of the places in which pendulum use is most widespread is France, where it was recently reported that over 7,500 noted physiologists use the pendulum on a daily basis for diagnostic and therapeutic purposes.

In everyday use, the power of the pendulum can supply precise and quickly available information that may well lead to a contented, healthy, and easier life. In whatever field comes to mind, you can find a reason to use the pendulum, as it considerably alleviates the difficulties of daily life. In business, romance, art, health, the home, and spiritual development, the pendulum can be an extremely useful tool, easy to use, and quick to provide results.

2

How do the pendulum and the divining rod work?

Until now, many different theories have been propounded about the mechanism of the pendulum and the divining rod. In my opinion, just as in defining our being, there are explanations on the physical, emotional, spiritual, electromagnetic, biochemical, and many other levels, which in essence explain the same thing in different terms, so various experts in the pendulum and radiesthaesia, as well as dowsers (people who use the pendulum and divining rod), also know the reasons why it works on various levels of understanding and awareness.

There are those who suggest that the pendulum works like radar. It locates various energies and informs the user of their location. This explanation makes sense in conjunction with the behavior of the divining rod or pendulum above mineral deposits or water sources, for instance, when the person uses the technique of standing near or over the place itself. Like real radar, the searcher sends a certain search signal, and it vibrates back through the pendulum or divining rod. Others believe that it is the target itself that sends the signal that vibrates into the pendulum or divining rod when the searcher stands above the objective. These theories explain the field search activity using the pendulum or divining rod, searches which are conducted in order to find water sources, veins of minerals, metal pipes in the earth, minerals, and other buried objectives.

But those theories do not explain the other varied

capabilities of these devices, such as sending a telepathic message using the pendulum, reading a person's character through a photograph, romantic readings, and the like, when there is no physical or tangible objective present.

The way that the pendulum works for informative yes/no readings, searching over a map, and other uses that do not involve a physical objective, can be explained by the hologram theory. A hologram is a picture which is in fact two-dimensional, but causes two-dimensional images to appear three-dimensional. The amazing uniqueness of a hologram lies in the fact that while, for example, you take a negative of a picture of a ball and cut a bit off, you will have less of the ball - or a flawed image of a ball. However, with a negative of a hologram, if you cut off a little of the picture of the ball, and beam a laser through the part you cut, you will still see a whole ball! According to this concept, the part contains within it the whole.

Thus, the hologramic, or holistic, idea was conceived, claiming that each and every one of us is in fact a part of one great and perfect whole which contains the entire universe, and that the entire universe is reflected within each of us. Hence, when we seek to find an answer to or information about something that isn't physically in our proximity, all of the information is already contained inside us, since we are part of the cosmic hologram that contains a picture of the entire universe; and by using the pendulum, we bring that information to our more conscious levels. The hologramic concept has ancient mystic and occult parallels, which contend that all human souls, including souls that are reincarnated as animals, plants, and inorganic matter, as well as the soul of various parts of the universe, in fact comprise a single whole. According to these concepts, your neighbor, the cat in the yard, the ground upon which you tread, and the bottle you hold in your hand, are interconnected by virtue of their being various aspects and parts of one whole, and are eternally interconnected. These ancient concepts were further validated when the atom and quantum theory were discovered. The famous biblical injunction to "love thy neighbor as thyself" is revealed in its full depth and glory in light of these neo-ancient discoveries.

An additional theory, which does not contradict the above theories, but rather ties in with them perfectly, is the explanation of the divining rod and pendulum as working through the presence of electromagnetic fields. Every object or organism in the universe is surrounded by a unique electromagnetic field. This field can be positive or negative, perfect or flawed. Organisms in the universe live in a perpetual state of energetic give-and-take between different energy fields, and they must preserve their own electromagnetic fields against harm from negative or energy-consuming electromagnetic fields. Numerous studies have shown how plants, for example, wither when negative or hostile energy is found close to their electromagnetic fields. Likewise, humans are likely to sense the presence of non-positive energy in various places, or emanating from different people or objects. You certainly must have found yourself in situations where for no obvious or tangible reason, you entered a room and felt discomfort or unpleasantness, or you spoke to somebody and afterwards felt exhausted and drained of energy without being able to explain the reason why. The place within you which senses these energies is transmitting its message to your nervous system and influencing your general feeling, with the aim of making you move away from that person or place whose energies are not right for you or are likely to harm you in some way. Every person has the ability to sense electromagnetic fields, and even to see various levels of those fields, which are called the aura, if s/he perseveres in developing this ability. Our nervous system, which receives these energetic transmissions unceasingly, enables us to receive any information we desire. A person with outstanding sensitivity for receiving energies sometimes doesn't need to seek an answer at all - it's enough for him to think of the question, and the answer will pop up on its own, since it is always within reach of our nervous systems, which are receiving the cosmic electromagnetic fields. But achieving this level of reception sensitivity involves a lot of practice, or an inborn talent (it's possible for everyone, however!), and thus we need various aids in order to decode the messages our nervous system is trying to send us. And this is, in essence, the role of the pendulum and the divining rod.

Thus, it is not the pendulum itself or the divining rod that gives the information, but rather each person's Higher Self which retrieves the information from the subconscious and brings it to the surface, the conscious, through use of the pendulum or divining rod. In other words, when the person using a pendulum or divining rod holds these tools above a given object or person, he is in fact measuring the interactions between a certain electromagnetic field and his own nervous system. But these interactions have no restrictions of time or distance. Just like you can cause yourself to feel sensations similar in strength to a given incident, if you bring it to mind and concentrate on it (as do some people who recall old fights and arguments, for example, and relive them), so you can sense energies which are at a great distance from you physically. This is how various methods of readings using the pendulum work, even from afar. Some claim that it is the subconscious picking up on the information and insights, while others contend that it is the superconscious or the Higher Self that receives these transmissions.

Another theory explains teleradiesthaesia (finding things out from afar) by stating that the conscious employs a certain kind of receiving and transmitting waves (similar to a radio or television set). A well-practiced person can direct his consciousness forcefully towards a given object, person, or idea, and tune himself to its unique frequency - somewhat like tuning a radio. Each frequency has its own speed and manner of resonance, and when tuning is achieved, nerve cells begin to resonate in response to the frequency being received. This resonance has its own frequency speed and wave length that give it its unique form, property, color, etc. The nervous system transmits this property and causes a certain movement that shows up in the pendulum or divining rod. Similar to the example of a TV set or radio, whose reception or transmission ability is harmed during interference of any sort (such as a storm, lightning, etc.), so also the human being's capability for reception is harmed when there are storms of thought or emotion.

It can be seen that the theories concerning the workings of the pendulum and divining rod are numerous, but all told they help the intellect understand phenomena that are beyond comprehension. Our logic is constructed according to thought patterns that we ourselves have created, and as such, they have only a partial capacity for understanding reality. Today it is known that reason, the mind, and the human intellect are capable of comprehending only a tiny part of all the phenomena present in nature and the universe. Hence, in the final analysis, it does not matter what the explanation for the mechanism of the pendulum and divining rod is; what is important is that they do indeed work amazingly well, and astound us again and again with their wonderful capabilities in all fields of life.

3

How to use the pendulum Methods for tuning and use of the pendulum

We must remember that the pendulum, by nature, is a device that receives nearly imperceptible resonances, reverberations, or radiation of which the person is not aware via his five senses, but rather only via his clairvoyance, which is used for predicting the future; and the pendulum responds to the radiation and reverberations through movement of its own. Because each occurrence of radiation is different and unique, the pendulum will respond in a unique way to each one. In other words, the pendulum will respond with a certain movement to the radiation of one person, and with a different movement to the radiation of someone else. Obviously, the number of its responses is very limited, but even so, there are over two hundred different kinds of pendulum movements.

A great deal of use is made of the pendulum for foretelling the future - or, put differently, to indicate a solution or the right direction toward the solving of any problem. There is nothing wrong with this use of the pendulum when it is the result of pure intentions; furthermore, a pendulum that is finely tuned by a trained person could refuse to provide an answer to a question about the future, for the person's own good. (For example, a negative answer to the question: "Will I marry John?" could end a relationship which is not meant to end yet. In such a case the pendulum is likely to not respond at all, even with repeated attempts!) But when the pendulum falls into the hands of charlatans who are not proficient in their craft, its use is dangerous, because people tend to believe and behave according to the pendulum's advice. Misuse of the pendulum will, of course, distort or falsify its responses.

It should be remembered while using the pendulum that it is being held by its user (or by the querent himself), as if it were part of the user . In order for the pendulum to indicate the right direction and receive the right energy, the consciousness of the person using the pendulum, and that of the querent as well, must be pure and free of obstacles and of outside influences.

Let me explain. When a certain method of guessing lies in the hands of the guesser, he will always obtain the answer he wanted from the outset. When a young girl plucks the petals off a daisy, saying: "He loves me, he loves me not," she will always find a way to get the desired answer, which she anticipated or hoped to receive from the beginning! This means that just as the pendulum can draw on the absolute knowledge that is within us to provide the correct answer to the question through its connection to the overall reservoir of universal knowledge, so our will, which has strong energy of its own, can transmit the desired information, even if it is incorrect, through its vibrations, and hence affect the movements of the pendulum.

In such a situation, the use of the pendulum is worthless. In order to receive correct, complete answers from the pendulum, its movement must not be diverted by perverse or distorted thought. Unfortunately, people who ignore this basic and fundamental rule often make use of the pendulum.

In principle, anyone is capable of operating and using a pendulum, on condition that he is a person who is sensitive to the cosmos in which he lives, is open-minded, has inner peace and a balanced mind (and body), and is capable of emptying his head of all thoughts (similar to deep relaxation). In other words, there are few people who are unable to use the pendulum. Obviously, its use requires patient and thorough study of the pendulum's principles of movement.

The first stage is for the person to test himself to see whether he is capable of being "joined" to the pendulum. For this purpose, we need to carry out the following experiment:

Place a large table in the middle of a quiet room, insulated against noise. Put the pendulum on the table. (A pendulum can be purchased at many shops, or you can prepare it yourself.)

Reach a state of calm by means of deep relaxation, balancing your breathing, or emptying your mind of thought - whatever you prefer. Be sure to relax your muscles, particularly those of the arm, shoulders, and nape of the neck.

Take the pendulum in your dominant hand (the left hand for a left-handed person), and hold the string between the thumb and the index finger, with the wrist completely relaxed.

Steady the pendulum with your other hand, so that it is motionless, hanging about twenty centimetres below the palm of the dominant hand. Now focus your thoughts.

The first thing is to check the direction of the pendulum's movement for "yes" (affirmative answer) or "no" (negative answer). Concentrate on the question: "What is 'yes'?" The pendulum will begin to move. Note the direction of movement and wait for it to stop. Steady the pendulum and ask: "What is 'no'?" The pendulum will describe a movement in the opposite direction of "yes."

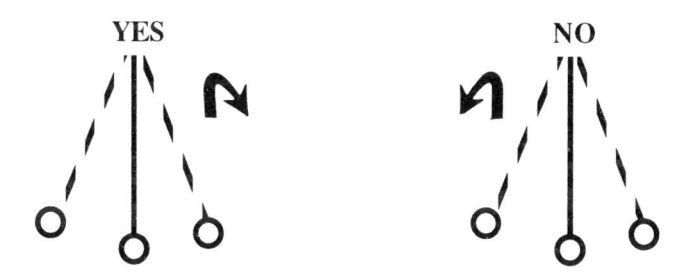

YES **NO**

If two opposite movements occurred, you are able to operate a pendulum!

Finding the "search position":

The search position is the position from which the user of the pendulum begins the reading or the search. This position helps the person connect with and tune in to the pendulum and receive permission to use it for a given purpose. This position is an individual one and can be a state of total immobility on the part of the pendulum, or a back and forth movement. In order to find the pendulum's search position, do the following:

Sit on a comfortable chair, be sure not to cross your legs or arms. (In other words, sit in an open position). Hold the string of the pendulum between your fingers. Look at it as though you expect to see a certain movement. Make sure that the fingertips holding the string are facing downward.

Now hold the string of the pendulum so that the pendulum is above your knees, exactly in the middle. Ask: "Show me your search position." Watch the pendulum. It is likely that the search position will be immobility, or gentle back-and-forth motions. In any case, this position shows that the pendulum is ready to receive and answer questions. If the pendulum moves in very large circular movements, backward and forward - change your position!

Methods for tuning the pendulum:

After you have succeeded in becoming connected to the pendulum, and after finding your search position, you will need to tune the pendulum so that you can recognize "yes" and "no" signals. There are many different methods, and each person, according to his feelings, will choose a particular way to tune his pendulum. Below are several tried and tested methods.

1. Self-tuning of the pendulum using power of thought

This method requires a powerful ability to focus one's power of thought and transmit it to inanimate objects. Although it requires some experience in working with power of thought, inexperienced people can also do it, so it's worth a try.

When you feel that you have "connected" with the pendulum in your hand, and you are relaxed, empty of thoughts, and balanced, you can tune the pendulum according to the codes that you determine. Sit next to an empty table in a quiet room with a calm atmosphere. Be sure not to cross your arms or legs. Hold the pendulum in your hand, gently steadying it with your other hand. Now, connect with it and focus all your thoughts on it, sending your message to it peacefully and lovingly, according to your inner feeling. For example, a clockwise circular movement can mean "yes" (positive answer); an anti-clockwise circular motion, for instance, could mean "no" (negative answer). In this tuning, when the person already has a certain amount of experience in spiritual work, and a good intuitive ability to connect with the pendulum, he decides which movements the pendulum will make to indicate a given answer. You will be surprised to find that the pendulum (quartz crystal ones are exceptionally good at this) will automatically make the movement you decided on, at the very instant you thought about it! So obviously, you must be decisive and clear about the movements it should make denoting a negative or positive answer, so as not to confuse the pendulum. If the pendulum makes the movement you were thinking about, it is a sign that there is a connection between the operator and the tool, and it is ready to serve you - for positive and pure purposes only, of course.

2. Tuning the pendulum by looking at its movements

In this method, it is the pendulum that tells you which movements are "yes" and which are "no." Some people prefer this method because the operator's objective state allows the pendulum to determine its own movements. This method is applied in exactly the way you tested your ability to be connected to the pendulum: after emptying yourself of thoughts and emotions, relaxing your muscles, and taking several slow deep breaths, hold the pendulum and ask it: "What is the pendulum's movement for 'no'?" It is important that you obtain clear and smooth movements. If for any reason you feel the pendulum is not answering you, you are probably tired or not concentrating adequately. You should rest a bit and then try again when you are refreshed and more focused.

The second stage after checking the pendulum's movements is recording clearly what the movements mean on a piece of paper. Apart from arranging things in one's head by writing them down, things will be better absorbed in your subconscious, and these signs become recognized signals between you and the pendulum. This is what you write:

The pendulum's rules of movement:
a. A circular, anti-clockwise motion from left to right means ....
(positive / negative).
b. A circular, clockwise motion from right to left means ....
(positive / negative).

Besides the movements whose significance you have determined, the pendulum is likely to make additional movements.

When the motion is elliptical (that is, not a completely circular motion), its meaning is similar to that of a circular motion (either 'a' or 'b'), but the answer is a weaker one!

A linear north-south movement or south-north, if this isn't your search position, is likely to be a negative answer. For maximal precision, the meaning of this movement should be checked with the pendulum itself, that is to say, ask the pendulum whether this movement signifies "no." In conscious tuning, some will program the pendulum to perform this motion as "perhaps not" or "perhaps so," that is, "yes" or "no" with a weaker significance - a possibility.

A confused diagonal movement which is not clearly defined attests to a disturbance in radiation, an interfering influence. The question should be repeated after taking a break.

If the pendulum stops or jerks in an erratic direction, it attests to a disturbance in radiation. You should stop for a longer time than in the case of confused diagonal movement. Sometimes this motion means that a question was not asked in the right manner, or should not be asked at all, or that it should not be answered at this stage. After you have learned the rules for working with the pendulum and have acted accordingly, you will avoid situations wherein the pendulum fails to answer as a result of an inappropriate question.

As a rule of thumb, you must watch out for circular movements, and try to distinguish between "yes" and "no," as this is likely to differ from person to person, and from pendulum to pendulum (as operated by the same person).

Now you should ask a question and interpret the pendulum's answer according to your directions of movement for "yes" and "no." Remember: The directions of the response can differ from one person to another, but it will always be a circular motion and will always be in opposite directions for positive and negative.

N.B.: Most men will find that the pendulum's anti-clockwise movement is an affirmative and its clockwise movement a negative answer; women will generally discover that clockwise movement is positive, and anti-clockwise is negative. Hence the following rules of directions must be adjusted according to the test you carried out!

3. Programming the pendulum by subconscious programming

The pendulum is a method of communication between the conscious and the subconscious. The subconscious can do nothing on its own except act according to the suggestions given it from outside - in this case, the operator's conscious mind. The operator must "teach" his subconscious which recognized signals or movements indicate "yes," "no," and "perhaps." The subconscious must fully comprehend this in order to establish correct communication.

The method of tuning the pendulum by programming the subconscious is excellent, since in addition to the fact that the tuning is precise and unequivocal, it also enhances the connection between the person and the instrument - the pendulum - and provides him with exceptionally good training to operate the pendulum for even higher and more complicated uses. This method is actually a harmonious and efficient combination of the two methods described earlier.

Take a blank sheet of paper and draw the following four symbols on it: A vertical arrow, a horizontal arrow, a clockwise semicircle, and an anti-clockwise semicircle.

Hold the pendulum above the vertical arrow. The string should be about 7.5 cm long at the beginning (remember, there is no "right" length for a pendulum string. You will gradually feel and find the length that is comfortable and right for you.) When you hold the pendulum, it will either be still or move randomly above the drawn arrow.

Now look at the pendulum. Using power of thought, cause the pendulum to move up and down along the direction of the arrow. Do not move it with your hand or fingers, but with power of thought and will alone!

Now hold the pendulum over the horizontal arrow. Again, look at the pendulum and use power of thought to move it along the arrow to create a horizontal movement.

Repeat the process, holding the pendulum over the clockwise semicircle, and, using power of thought, make it move in a circular motion according to the direction of the circle.

Repeat the process with the anti-clockwise semicircle.

When you feel adequately confident with your ability to move the pendulum using power of thought, go on to the next step: Make the pendulum move clockwise in a smooth and complete circle while holding it over the clockwise semicircle. While the pendulum is still circling, talk to your subconscious aloud, calling it by name, and say: "When I ask a question and the answer is 'yes,' make the pendulum move clockwise, in the same direction it is moving now." Or say: "This movement means 'yes.'" Your voice should be steady, confident, and authoritative, so that your subconscious will obey your conscious precisely.

Now hold the pendulum over the anti-clockwise semicircle and use power of thought to make it move along the direction of the circle. While it is still circling, tell your subconscious: "When I ask a question and the answer is 'no,' make the pendulum move anti-clockwise, in the same direction it is moving now." Or say: "This movement means 'no.'"

Repeat the programming once a day for a week. This

repetition will ensure that your subconscious has absorbed your wishes fully.

Whether you have chosen the first or the third method, you will generally find that the pendulum obeys your thoughts. Most people are surprised when first seeing the pendulum move in accordance with the transmission of their thoughts. Its movements prove quite simply that thought has the ability to affect matter.

If you have encountered difficulty moving the pendulum with power of thought, it is probably because you are tired (this is the main cause of the pendulum failing to obey), or you are not capable of transmitting thoughts that are strong enough to cause movement. This situation is extremely rare. If you rest and then try again when you are calm and alert, and are still unable to move the pendulum using power of thought, you must begin building higher mental voltage and mental strength by doing this exercise every day. If you persevere, you will eventually cause the pendulum to move. Just like one can strengthen the body via physical exercise, so one can strengthen the brain and the thought force by mental calisthenics.

After you have programmed the pendulum to respond with "yes" and "no," you can also program it, or decipher its movements (according to the first technique), for the concepts of "affirmative/positive" and "negative," or "harmonious" and "disharmonious." If you program it yourself, it is best to use the movement signifying "yes" to signify "affirmative" or "positive" and "harmonious," and the movement signifying "no" to signify "negative" and "disharmonious." These concepts will enhance your use of the pendulum.

After you have tuned and programmed the pendulum, or have tuned in to it, there is no need to program it again. For this to be so, you must ensure that it remains in your possession, and prevent others from touching it, since extraneous vibrations can upset its programming. A pendulum which has been put aside and not used for a long time will probably need a bit of "refreshing" concerning its movements for the answers you tuned it for. If you have chosen to use a stone (quartz) pendulum, it is good to place it occasionally on a cleansing quartz crystal cluster, and afterward lightly program it. If someone else has used your stone pendulum (not a good idea), it is best to rinse it in running water and place it on a quartz cluster for two or more hours (according to how you feel).

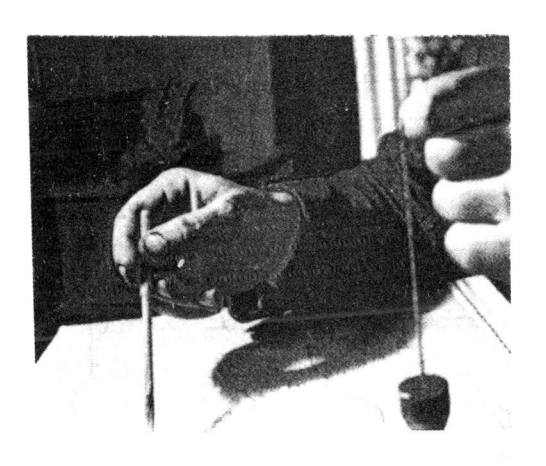

4

Rules for use of the pendulum
Tuning in to an objective - How to start

The more the pendulum operator succeeds in tuning in to the objective he is asking about or searching for, the better and more accurate the results will be. One of the most important processes in beginning the process of concentrating on the goal is the following: First of all, you must state to yourself what the objective is - the question, object, person, etc. in regard to which/whom you are going to use the pendulum. This statement must be understandable to the pendulum. To do this, hold the pendulum and declare your intention.

Afterwards, you must check whether you possess the skills to perform this reading or check. Sometimes there are questions that the person is not yet at an adequate level of consciousness to deal with. Therefore, after having stated your objective, hold the pendulum and ask: "Am I able to do this?"

Because there are certain areas of pendulum use that not everyone is permitted to enter into (such as various methods of foretelling the future, certain types of personal readings, and so on), the permission depends on the person's level of consciousness, his ability to rid himself of internal agendas, and various other criteria. In order to find out whether you are allowed to perform a check, reading, or any other particular use of the pendulum, hold the pendulum and ask: "Am I allowed to do this reading [check, or any other use]?" Or: "Am I allowed to ask this question?"

The last thing you must check is your readiness at this very moment to make this particular use of the pendulum. Many different factors can affect this readiness: for instance, fatigue, crossing legs or arms while doing the check, the presence of powerful electronic devices near the pendulum, lack of inner quiet, and more. In order to check that there are no interfering factors, or that you didn't overlook any stage involved in achieving the specific objective, hold the pendulum and ask: "Am I ready?" Or: "Am I properly attuned?" Or: "Is there anything I've forgotten?"

Asking these four questions before each use of the pendulum will ensure good and trustworthy results. If you receive an affirmative answer to all of the questions, continue using the pendulum for the objective you have set yourself. On the other hand, if one of the questions receives a negative answer, it is best to stop, put the pendulum aside, and try again another time. A negative answer to the question: "Am I allowed?" could be telling you explicitly that it is best for you not to venture into the area you wanted to check at all. If you insist and decide to continue in spite of the negative answers, be aware that you cannot rely totally on the pendulum's answers.

Apart from these initial rules, you should take into account several factors that can affect the movements of the pendulum and the pendulum's own capabilities:

The pendulum will answer any "yes/no" question. Thus you should formulate your questions so that the answers will require only "yes" or "no." Furthermore, the questions should be as precise as possible.

Your emotions and thoughts should be in a neutral state; you must not have an opinion or emotion regarding the answer, as such opinions, emotions, or desires are likely to affect the answer. You must want the genuine answer only, even if it is not what you want to hear. For this reason, you should practice various relaxation techniques, and tell your subconscious: "I am emptying myself of all emotions and thoughts. I wish for and am receiving a true answer only." This task is not an easy one, but after you have achieved the ability to be neutral and objective in your use of the pendulum, you will have acquired a wonderful and powerful tool that apart from being helpful in everyday life will develop your intuition and sixth sense. Until then, when you want to ask a question that is very important to you, it is worth getting help from a friend who knows how to use a pendulum, but who is completely neutral about the answer, and has no opinions or emotions about it.

While using the pendulum, you should be relaxed and serene. If you are agitated, disturbed, irritated, sad, or feeling any other oppressive or upsetting emotion, the reading is liable to fail. The pendulum may move jerkily, skip, remain immobile except for small nervous motions, and perform various movements that are likely to upset the reading.

During the reading, make sure that your hands or legs are not touching each other or folded and don't sit with your legs crossed, as these positions undermine the energy flow that is vital to the movements of the pendulum. Sit with your legs apart, feet planted firmly on the ground. Hold the pendulum in your right hand (or your left, if you are left-handed), with your other hand resting lightly at your side.

When you ask a question, you must understand its significance. Pay attention to the words you choose to use, and be aware of the essence of your question. Questions such as: "Will I be successful with the opposite sex?" "Will I be successful in life?" or "Will things be good for me with

Ryan?" are a common mistake among beginners. Many of the askers do not know exactly what the intention or significance of "success" is, even regarding themselves, so the use of abstract concepts leads to failure. The questions should be as precise as possible. If being successful in life for you means getting a particular job, that is the question you should be asking.

During the reading, pay attention that you aren't in an area where there are a lot electrical appliances that operate at high voltage such as X-ray machines, color TVs, computers, and so on. These devices emit powerful radiation that tends to upset the motion of the pendulum and make the reading difficult, especially when the reader is not experienced.

It is best to do the reading in as quiet and peaceful a place as possible. Sometimes, the numerous frequencies and resonances in noisy and crowded places are likely to affect the movements of the pendulum (making it move and skip about erratically).

5

Preparation of the pendulum

Choosing a pendulum

There are now many different types of pendulums on the market, ranging from simple ones made of plastic to ones made of gold and precious stones. When purchasing a pendulum, you should remember that monetary value is insignificant. A pendulum made of a button hung on a string is no less effective than a pendulum made out of expensive metals or stones. However, if you plan to use the pendulum for healing purposes, a crystal pendulum is strongly recommended because of the qualities of the stone itself; but again, it can be the very simplest, and any other simple pendulum will also be appropriate. In any case, whether you buy the simplest or the most sophisticated pendulum, the only thing you must make sure of is that you feel comfortable using it and feel a sort of "empathy" with it. You must also check that it isn't too heavy or too light for your taste when holding it. Of course, you can make your own pendulum. In this case, the "connection" between you and the pendulum will form automatically, since you prepared it yourself and invested your personal energies in it.

Making a pendulum yours

A pendulum can be made out of any material, though many prefer a pendulum made of gold or a crystal or a good stone. Many prefer a hollow crystal that can be filled with holy water or essential oil.

It is important that every pendulum have a centre of gravity which points clearly downward (a point that concentrates the weight to the lower center), and a ring to which the string will be tied for holding it. There are also special strings for preparing pendulums; these can be purchased.

Nowadays, since the use of pendulums has become common and available to many people, be they healers or mystics, or people using it to solve everyday problems, the manufacture of pendulums has become an industry. Producers of pendulums, including even those who are radiesthaesists themselves, work day and night in their workshops and laboratories to prepare new types of pendulums. At present, tens of thousands of different kinds of pendulums can be found on the open market. The most common materials that pendulums are made of include wood, metal, bakelite, plastic, amber, and various quartz stones. Pendulums made of expensive and exotic metals and materials such as gold, silver, beechwood, ivory, marble, Czech crystal, jade, amethyst, lapis lingua, whale bones, etc. are rarer.

The body of the pendulum

Most radiesthaesists and makers of pendulums claim that pendulums made of metal should be used only for specific purposes, because they respond very strongly to outside influences. This is because metals act as conductors. However, the metallic vibrations tend to interfere with and affect the precision of the reading and the reception of messages. An iron pendulum, for example, is considered exceedingly sensitive to magnetic fields. A copper pendulum is extremely sensitive to the absorption of electric surges. But non-conductive materials, such as glass, wood, and plastic, are considered to be very good for pendulums.

Pendulums are usually conical, spherical, spiral, cigar-shaped, or cylindrical. It is imperative for them to be symmetrical. Most users of pendulums prefer the cylindrical shape because of its precision. The spiral-shaped pendulum appears to have the advantage of being less affected by gusts of wind, but sometimes has trouble picking up the more subtle resonances that are ultimately very important. There are pendulums with a hollowed-out core in which some of the material being checked can be placed, as well as pendulums with a compartment into which holy water or aromatic oils can be poured. However, the combination of the additional substance, whose composition is not always known to the user with absolute precision (the substance can include additional ingredients that are undesirable for the specific reading with the pendulum) inside the pendulum and its composition, can affect the movements of the pendulum.

The string

The material being used to connect between the fingers of the user and the body of the pendulum must be flexible enough that the pendulum can balance itself freely. Thread, chain, nylon line (fishing line), black silk cord, various kinds of twine, wire, white hemp, or string are usually employed.

Besides flexibility, the string must be pliant and strong. Sewing thread cannot usually hold the weight of the pendulum and breaks within a few moments.

Preparing a pendulum at home

To make a pendulum, you should take a glass or

plastic bead and tie it to a piece of string, preferably black silk. A small ring (such as a wedding ring) can be used in place of the bead. Later on, when you become more experienced, you will have to take into account the metal the ring is made of, as well as the significance behind it and the energy with which it is imbued - wedding ring, friendship ring, etc. If you have chosen to make your first pendulum out of a ring and are unsure whether or not it carries solely positive and effective energies, it is best not to use it unless you can cleanse it of its previous energies by means of the power of thought. To be on the safe side, place it on an energetically cleansing crystal colony.

A button is also a common option for preparing a pendulum for beginners. Black thread is recommended because from a scientific viewpoint, black emits fewer interfering vibrations than do the colors of the spectrum, and therefore reduces the possibility of distracting interference. A piece of artificial pearl in the form of a shirt button can be used, tied to the end of a thick piece of cotton string about 20 cm long. A pendant-shaped earring of artificial pearl can also serve as a good pendulum for beginners. All you have to do is cut the catch off the earring and tie the ornament to a piece of string or a thin chain.

Other possibilities include sticking a needle into a ball of aluminum foil, cork, or any other substance the needle can pierce. The eye of the needle will make a convenient place to tie the string or chain. Obviously you must ensure that the point of the needle isn't sticking out, so as to avoid an accident.

An ornament, pendant, or fishing weight can also be used for making a pendulum for beginners. Yet another method is making a pendulum from three parts, which includes adding a small stick to the string and weight. The French radiesthaesist, Henri de France, suggests that beginners use a wooden ball or cylinder weighing 30-60 gm (one to two ounces), and a strong flexible cord about 25 cm long ending in a small loop. Make a hole in the ball and attach the loop to it with a small wooden peg. Tie the string to a small stick about 20 cm long and 0.8-1.7 cm in diameter, and tie it to the near end of the string. Henri de France explains that this type of pendulum should be held by the small stick between the thumb and index finger. When the user finds the length of string which is comfortable for him, the pendulum will begin to rotate in accordance with the vibrations emitted by the object for which it is being used.

An advantage to a pendulum you make yourself is that it is personal and unique, and is exactly in tune with your own resonances. Not only beginners choose to make their pendulums themselves; advanced radiesthaesists and healers also often prefer to make their own unique pendulum that carries their own vibrations and is appropriate for their purposes.

Tuning for preparation

In all instances, it is a good idea to carry out a special tuning before beginning work: Approach the work calmly and serenely, in a room with as clean and pleasant an atmosphere as possible (you can put on pleasant music and use aromatic oils), place all the materials on a work table which is free of extraneous items, relax your body, and clear your mind of all thoughts and emotions. It is best to request guidance and direction, and to use techniques that fill you with the energies you prefer. You should no begin work when you are feeling highly emotional, angry or sad, and you should check yourself to see that you are free of any negative intentions. Such energies can leave a residual impression on the pendulum that you are making, and can incorporate non-beneficial energies that will adversely affect its work and its accuracy.

6

The pendulum in romance

The pendulum has numerous uses, but one of its most popular uses, both by amateurs and by professional radiesthaesists, is romantic readings. The pendulum is one of the most successful and accurate instruments for measuring and checking the compatibility between two people. The advantage of the pendulum is that you do not have to be in the physical presence of the person in order to test his suitability. You can think about him, imagine him or tune in to him. In fact, you don't even have to know him personally or receive any information about him in order to check him. In such a case, the person's name will definitely be enough.

One of the theories that focuses on the pendulum's being operated by the subconscious explains the success of romantic readings using a pendulum as being the result of receiving the electromagnetic fields of the people involved. When two people have incompatible electromagnetic fields, short circuits are created between them, the speed and force of which depend on the extent of the incompatibility between their magnetic fields. The significance of the magnetic field in human relations is seen in those familiar situations in which we feel, for example, an inexplicable aversion and repulsion from a person we've only just met and who has done nothing in reality to arouse those feelings toward him. The reason for this is that hostility or opposition is present in this person's electromagnetic field, or that his magnetic field (aura) is totally incompatible with our own. But when the relationship of a couple is in question, the importance of the capacity of their energy fields to merge assumes inestimable importance. When there is harmony between two magnetic fields, the chances of establishing a fruitful, loving, stable and genuine relationship are very great.

Checking the compatibility of a couple

The technique for checking the compatibility of a couple is very simple. You can do it yourself, but you must be sure you are capable of remaining neutral. If you would like to check whether the love of your life is in fact compatible with you, you can carry out this simple test. If you are having trouble choosing between several romantic interests, you can even check several people with this test.

Sit in a quiet, airy room with a pleasant and calm atmosphere. Make sure there are no high-voltage electrical appliances in the room. Place a sheet of paper in front of you, preferably on a clean and level surface, or on a table that is free of unrelated objects. On the piece of paper, write your name and the name of the person you wish to check. For example:

Ron (the checker) / Gillian

If you want to check four additional people, write their names:

Anna / Vivian / Shelly / Diana

Sit on a comfortable chair near the paper. (Remember not to cross your legs or arms - this interferes with the test.) Make sure you are alert and calm. Relax your muscles a bit. Empty your mind of any opinions, thoughts, or emotions, and get your consciousness into a neutral state. (Remember!

If you cannot neutralize your emotions, they are likely to detract from the accuracy of the reading!) Take several deep, slow, restful breaths, and feel yourself relaxed and calm. State the purpose of the check, hold the pendulum, and ask the opening questions ("Can I carry out this test? Am I allowed? Am I ready?") Hold the pendulum over the two names and ask: "Are these two people compatible?" When the two people are truly compatible, you will receive a smooth, obvious "yes" movement (the movement signifying "yes" according to how you programmed your pendulum). If they are not compatible, you will receive a "no" movement. You can measure the degree of compatibility - or lack of it - by observing the movement of the pendulum and measuring the force of its movement. A faster and broader movement, a gyration with a larger diameter, will enhance the answer, negative or positive. Sometimes a situation occurs in which the reading is performed under all the proper conditions (see above, in the chapter on Rules for Use of the Pendulum), but the pendulum still does not make any significant movement, but rather small, jumpy, trembling, or skipping movements. If you are extremely sensitive, you can sense a sort of light feeling of numbness or pins and needles in your hand. This feeling means that you have tuned into a conflict between two exceptionally unharmonious energy fields. A relationship between these two people - and particularly a romantic relationship - will be destructive and fraught with defective communication. Therefore, it is definitely not recommended.

Evelyn's story

I first heard of Evelyn from the many stories told by her elderly mother who lived in my neighborhood. Her mother would tell marvelous stories about her - her success, her beauty, her talents. I understood from her mother that she was a very ambitious girl. Indeed, her advancement in her career was meteoric. At age 32, she had been made the manager of a branch of a large bank, and derived great success, esteem, and satisfaction from her work. The only thing clouding her mother's happiness was that even now, at age 34, she had not yet settled into a stable relationship. The elderly mother would express her pain to me. She couldn't believe how her "perfect" daughter could not find a husband and start a family.

One day, when she held a birthday party in her garden, I met Evelyn face to face. Indeed, all the proud mother's stories about her beauty and intelligence turned out to be true. She was a beautiful girl, slim and shapely, with a long auburn mane of hair, well-groomed and impressive. It was impossible to miss her. When we sat down at the table, enjoying my neighbor's wonderful garden, we began to talk, and in spite of the obvious difference in our views of the world, we quickly became good friends. Evelyn herself brought up the matter of her single status. She said that she had long since learned to deal with the oppressive family pressure and that even though she felt a small twinge in her heart whenever they asked her yet again: "Well, have you found anybody?", she would make it clear to those she knew that she felt perfectly fine about being on her own. Her main excuse was that she did not have time for relationships. She was in fact a very busy woman, but the real reason, as she revealed to me, why she was unsuccessful at sustaining a relationship for more than two or three weeks was her choosiness and inability to compromise. Her self-esteem was extremely high, and she sought a man who would fulfill her criteria. "Look," she told me, "I'm not short of money, and I don't lack interest in life. My work is fascinating, and the salary is more than enough for all my needs. The man I'm looking for has to be successful in every area, a man I can admire both physically and mentally, a person of stature." Unfortunately for her, she did manage to meet several such men - pleasant, intelligent, interesting, and highly placed. But within a short time, it became apparent that they weren't interested in relationships, weren't ready for marriage, or were unable to create emotional ties and enter into a deep and committed relationship. Her disappointment in males was great, but she was totally unwilling to compromise.

One evening, my neighbor knocked on the door. She came in with a desperate expression on her face. "I don't know what to do with her," she said sadly. "I tried to introduce her to a wonderful man, the son of a childhood friend. He's a talented doctor, a delightful and charismatic man, and when I saw him two weeks ago, I knew straight away that he was the right guy for her. But much as I've tried to persuade her, she isn't willing to listen. She gets angry when I try to introduce her to anyone, and has told me to stop interfering in her life. You know that I try not to interfere or to pressure her, but I have the feeling that this is the man for her. Maybe you could try and speak to her?" I explained to the mother what my approach to relationships was, and suggested that before she continued driving Evelyn crazy, I would perform a little check of their names with a pendulum. I rang up Evelyn and requested her permission to carry out the check in her mother's presence. Evelyn naively thought that this was a clever trick to placate her nervous mother, and to give her a bit of respite from the unceasing entreaties to meet him. When I asked if she would like to participate in the test, she replied that with all due respect, she didn't believe in such nonsense.

I asked the worried mother to calm down a bit because I could sense the interfering energies of anxiety. I sat down by the table and wrote the names of Evelyn and the man on a sheet of paper. I took a few deep breaths, emptied my mind of all thoughts and feelings, and held the pendulum above the names on the sheet. In response to my question, "Are these two people compatible with each other?" the pendulum described a smooth, clear, clockwise circular movement, which indicated an affirmative response. The diameter of the circle was relatively large, and it rotated rapidly and confidently. Thus, according to the initial reading, the two were suited to each other. I continued to check the following criteria: physical, emotional, mental, and spiritual compatibility. The answer was amazing: the pendulum answered every question with a clear and energetic "yes." When asked the question, "Are the couple suited to each other spiritually?" the pendulum described a clockwise circle with a particularly large diameter.

I hurried to ring Evelyn with the results of the reading. I explained to her briefly how the pendulum works, and gave her various examples of the pendulum's success and help in the past. She refused to believe it. Finally, after grasping the more scientific side of the pendulum's action, how it could pick up on and tell us about the electromagnetic field that surrounds every person, and indicate the compatibility between those fields, she said she must see it for herself. Within half an hour she showed up at my flat, with a doubtful

Pendulums

look in her eyes, but I could see that in fact some interest had been aroused in her. Of course, the pendulum answered the questions with the same precision as before. (It should be noted that because Evelyn approached the reading with almost objective interest, not attaching much importance to the pendulum's answers, yet overcoming her skepticism and letting her natural curiosity take hold, her feelings did not interfere with my reading, and the pendulum responded as well as before.) The fact that the second reading was identical to the first amazed her. Her analytical mind knew that the chances of the pendulum repeating the same movements precisely by coincidence were negligible. In the end, she agreed to take the man's telephone number and speak to him to see what it was all about. Three weeks later, she rang me in excitement. She didn't want to jump the gun, but she had liked him from the first moment, and the first date, as well as the many following it, was simply wonderful. A year and a half later, I got to see her at the altar, glowing with happiness. Today, three years after the reading, the couple have their first child, and enjoy a wonderful relationship.

7

The pendulum in work and career

Choosing a career is one of the most important choices a person has to make. When a person chooses a career that does not suit him, it doesn't matter how good he is, how quickly he advances, or how much money he makes: there will always be a part of him that remains unsatisfied. This is because each of us has a certain vocation in life, and only if it is realized will we feel satisfied and completely content.

In the distant past, before civilization took over, humans were closer to nature. This closeness to nature, a more natural lifestyle, and simpler aspirations, made it easier for people to identify their vocation. When they still lived in tribes, some children naturally took a great interest in accompanying the warriors, hunting alongside them, and observing their actions. Other children spent time observing the work of the toolmaker, imitating his actions and trying to make a bone tool themselves. A child who felt a strong urge to draw would spend his time trying out various materials to create colors, while a child who was fascinated by the works of the tribe's healer and wished to imitate him would join him, serve him, and learn his techniques over the years. These children would naturally find the career paths that realized their purpose in this world.

Today, many parents tend to force their children to learn "successful," "lucrative" professions - ones that will provide them with high status and wealth. Some professions are considered to be more popular in the "market," and many young people go to study them, when in fact they do not really suit the inclination of their hearts, and are not the genuine realization of their vocation. Sometimes, when the person realizes this, he feels it is already too late, that he is too old and tired to learn a new profession. In actuality, this is not true, and the person can realize his true vocation at any age and in any situation if he really wants to with all his heart. But why wait for so long? In these times, when success is measured by money, property, and often fame, many people are forfeiting their true vocation in favor of jobs that do in fact bring in a lot of money, but which oppress them spiritually and make their lives impoverished, boring, and unsatisfying. Al's story is a good example of a person who, at a mature age, was helped by the pendulum to make a career change and bring new light into his life.

Al was a major liquor importer. He was very successful at his work, and at age 38 was already the owner of a fancy house in an exclusive neighborhood, and had a loving wife and three children. But he felt completely dissatisfied with his life. He believed he had reached the "mid-life crisis." Nothing interested him anymore. He began displaying impatience and irritation both at work and at home. From being friendly and gregarious, he became an almost impossible person, even in his own eyes. He signed up at a gym, participated in marathons, had several extra-marital affairs, went out a lot to bars and restaurants, and drank alcohol regularly. He began to feel depressed and lifeless, and nothing cheered him up or excited his interest. He felt that his life was going downhill and knew he had to do something. He was referred to me by a mutual friend. It was hard for me to put my finger on the reason for his depression. From what he said, his lack of enthusiasm for his work and the way that he spoke about it, I decided it was worth trying that direction. I asked if he'd thought about a change of career. He said that although he was becoming increasingly fed up with his work, he knew that many people would be glad of such a fruitful and prosperous career as his. I suggested that we try and focus on this issue. I used a pendulum reading and a process of elimination (discussed below). The amazing result was that of all the careers I asked Al to list, "tour guide" was the profession the pendulum indicated. At first, Al couldn't understand why that was the choice. After we had repeated the process several times, he began to remember why in fact he had begun dealing in import: What fascinated him, in fact, was not the financial profit in import, but the thought that a career of this type would allow him to travel to faraway places and visit new countries. Over the years, apparently, his love of travel, history, and stories of those faraway places that he used to love so much in his childhood and youth was extinguished, and he forgot and repressed his true desire. After several meetings, Al decided to try and check out this new field. He took several courses on tourism. Very quickly his life was filled with new light. He found that the studies fascinated him, and within a few years, he began doing tour-guiding in addition to his import business. He was excited to discover that working with people, the ability to inspire enthusiasm by means of the extensive knowledge he had acquired and to direct their attention to the beauty of the places in which they were traveling, thrilled him and caused him to exploit his talents to the full. In parallel, he felt that his depression and boredom were disappearing, and that he was not only going back to being the same calm and cheerful Al as before, but really felt that way inside. Of course, the change of career affected all areas of his life, and his family life improved amazingly.

Find yourself a quiet, calm place in which to sit. Sit comfortably and be relaxed and peaceful. Take quiet, slow, deep breaths. Empty your mind of any thoughts, worries, emotions, or opinions. Remember, you are interested in finding only your true purpose. You must be refreshed and alert during the test, so it is best that you carry it out during the hours when your energy is at its peak. If you are tired or pressured, you should rest for long enough to become refreshed and calm before beginning the reading.

Declare the objective of the test and ask the opening questions. Now start going over each profession on the list, one at a time, with the pendulum. While holding the pendulum over a profession, ask: "Is this profession the right choice of a career for me?" Many items drop out as a result of the pendulum's negative response. At the end of the first reading, the pendulum will probably have indicated "yes" for several professions. Check those out on another day. Continue this way until only one or two professions remain. This is likely to take several days. The professions which were selected are the fields in which you will apparently achieve the most success and satisfaction.

Choosing an appropriate job

If you have several job proposals before you and are not sure which of them to choose, write them down and pass the pendulum over each one in turn, asking: "Is this the job that is appropriate for me at this time?" If you received a positive or negative answer for some or all of them, ask: "Is this job better for me than the other ones?" One place will produce the smoothest, clearest positive movement of the pendulum. If you continue receiving negative responses for all the jobs, most likely all these workplaces contain factors that are not positive for you. Don't forget that a place of work is not measured only by the type of profession or the salary, but also by the work environment, the type of relations between the employees, the nature of the management, the atmosphere in the place, promotion possibilities, and any other factor you consider important. It is probable that a glittering position that received a negative response from the pendulum lacks one or more factors that have conscious or unconscious significance for you.

8

Using the pendulum for diagnosis and healing

One of the uses of the pendulum most common among professional pendulum experts is for medical purposes. Many of the pioneers of the use of the pendulum were physicians and physiologists. In France itself today, there are more than 7,500 physicians and physiologists who used the pendulum as an inseparable part of their work, similar to the stethoscope, and many scientists are also researching its use in the field of medicine.

Using techniques for checking foods, every person can find the exact types of foods that are beneficial to his health and that of his family. These techniques are so simple that some of them can even be applied at the supermarket or mall. It is worth performing the tests in the presence of all the members of the family in order to discover which foods are suitable and healthy for each one. These foods are likely to differ from person to person, even within the family itself.

First method of checking the foods according to a list

The foods should be divided into groups according to any classification you choose. There's no need, nor would it be particularly easy, to include all the foods you can think of in the list. It is enough to include all the foods that are usually eaten by members of your family - the foods they like, and foods they usually eat on special occasions. But it is important to include all the foods that are eaten regularly by the members of the family, and it is important to include the foods that are known to be healthy, but are not common in your household, such as wholewheat bread, brown rice, saltwater fish, and the like, because some members of the family could discover that it happens to be those foods that are most vital for them. The list should be divided into groups according to criteria of some sort - no matter what, but the groups must be organized and clear (according to the food groups, the type of food, and so on). Each group should be written on a separate sheet, with the name of the person for whom the food is being checked written at the top of the page. After preparing the lists, the pendulum should be passed over each name, with the operator asking: "Is this food healthy and right for me [or for the person for whom the foods are being checked]?" An affirmative answer from the pendulum will indicate that the food is beneficial to the person. A particularly strong positive response indicates the possibility that the person needs a given food in a relatively larger quantity than other foods on his menu. In contrast, a strong negative answer indicates that the food is harmful or undesirable for the person.

Second method: Checking the foods themselves

This method is more efficient primarily when checking food products that are complex, ready-to-serve, or purchased, such as a pie containing a number of ingredients or a snack food comprising various food groups and additives, and the like. In order to perform the check, place the food being checked on a table cleared of any other objects and food items. After tuning and asking the opening questions, hold the pendulum above the food and ask: "Is this food appropriate and correct for me [or for any other person for whom you are carrying out a check]?" Watch the pendulum's movements. A positive answer by the pendulum will tell you that the food is suitable and healthy for that person. The advantage of this check is that it can already be performed at the supermarket before you purchase foods that may prove to be unsuitable. Of course, you will probably get a few astonished glances from other shoppers. Don't hesitate to explain to them what you're doing if they ask. The greater the number of people who get to know the advantages of using a pendulum and the importance of eating a healthy, correct, and nourishing diet, the greater the contribution to the health of society will be.

Third method: Checking diet menus

If you have a specific diet recommended by a dietician or a doctor, with a list of foods and the quantities of each food you're supposed to eat, it is worth your while to check if that diet indeed suits you as much as possible. To do this, write your name at the top of the list, or if it contains text or diagrams that have nothing to do with the foods being checked, copy it onto a clean sheet of paper. If the quantities that you are supposed to eat appear next to each food, copy the types of the foods onto a separate list, so that you will then have two lists - one containing the names of the foods only, and another containing the names of the foods and the quantities that you are meant to eat of each item. At the top of each list, write your name or the name of the person for whom you are performing the check.

Discovering shifting and dislocation of the vertebrae in the spinal column

The importance of a healthy and stable spinal column for general health is known to everyone. The condition of the spinal column affects the body's general stability and the structure of the entire skeleton, and consequently the health of the internal organs as well. The spinal cord, which passes through the vertebrae and out of nerves that extend to all the parts of the body, is directly affected by the condition of the spinal column and its vertebrae. The electrical pulses from the brain to the rest of the body parts pass through the spinal cord, and movement, however slight, of one of the vertebrae will cause pressure on the spinal cord and the adjacent vertebrae, and will block the flow of nerve energy required to operate the muscles. In such a case, the situation is likely to gradually deteriorate to the point of pain radiating to various organs of the body. In serious cases, this can lead to an inflammation, a slipped disk, and paralysis. These movements can be corrected efficiently and quickly by an experienced chiropractor, but they must be discovered at an early stage before the situation has deteriorated and become serious.

In order to locate movement of vertebrae at an early stage, many doctors and chiropractors use a device called an analagraph. The analagraph is a machine which passes over the skin along the length of the spine and picks up any emission of unusual heat on the spinal column. These points are noted on special paper and printed in the machine, so that the chiropractor can look at it and identify the places in which there are spots of abnormal heat on the spinal column. These spots indicate the incorrect positioning of vertebrae in the spine. But the main problem with this device is its costliness. In contrast, using the pendulum to diagnose vertebral shifting is no less accurate, and in many instances even exceeds the precision of the analagraph, for the simple reason that the pendulum is capable of picking up on the frequencies and vibrations of a shifting taking place even at the earliest stages when the situation still appears to be normal. This advantage derives from the reason that before a given situation of physical imbalance is created, that imbalance shows up in the body's electromagnetic fields.

In order to make the matter clearer, imagine that a certain person is in a situation in which he takes upon himself a heavy workload, and, in addition, he bears the responsibility for his wife and children, for his elderly parents, and for a close friend who is in deep financial difficulties. In short, he's carrying the whole world on his shoulders. This emotional state would, on the face of it, later express itself in the condition of his spinal column. The great pressure, the burden, and the thought pattern that idealizes taking exaggerated responsibility for his surroundings, will soon show up in the condition of the vertebrae in his back. At the moment, he does not feel any unusual pains. However, the muscles of his shoulders and back are frequently tight and hard (which directly affects the condition of the vertebrae!), but he's already gotten "used" to that and doesn't pay any attention to it. In actual fact, the lack of balance in his spinal column already exists. Now, all he needs to do is lift one medium-sized suitcase using the wrong movement, or take upon himself one more small emotional responsibility, and a slipped disk will lay him up for a long while, when in fact it derives from the body's defense mechanism, which sent him repeated messages that the load and the burn-out were too much. When a person does not listen to his body's warnings, the body has its own ways of getting his attention and putting an abrupt stop to his unbalanced lifestyle.

It is therefore important to locate these slight and barely perceptible movements of the vertebrae in order to reach the correct conclusions. The person will probably need a chiropractor to put the vertebrae back into place. Deterioration of the condition will probably be halted. The body will gradually bring itself back into balance through rest, a change in lifestyle and thought patterns, Bach Flower remedies, massage, Shiatsu, and numerous other efficient methods. But first of all, we will attend to detecting the movement of vertebrae and their condition along the spinal column:

Declare to yourself the objective of the check (examining the condition of the patient's vertebrae and discovering any movement). Ask if you are able to do this, if you are allowed to, and if you are ready. After you have received permission to begin working, take several deep, slow, comfortable breaths, empty your mind of any emotions and thoughts, and ask for help from the Universal Force (or whatever other name you call it by) for your work. (Requesting assistance is always important, especially when we want to diagnose and treat, and when we are working on the energetic planes, as it helps us to neutralize our ego whose vibrations are likely to interfere and be disruptive. It protects us from receiving non-positive energies from the patient, and helps us in many other ways that will not be discussed here.)

Now ask the patient to lie on his stomach on the floor, on a mattress, or on a treatment cot. It is important that the patient lie in a way that his back and neck are straight and relaxed. Ask him to breathe slow, deep, comfortable breaths so that he will be relaxed. Begin the examination from the point of the foramen magnum at the base of the skull (the depression at the base of the skull from which the cervical vertebrae extend), and continue passing the pendulum along the whole length of the spinal column down to the last vertebra of the coccyx. Hold the pendulum above each vertebra; you can touch it with your free fingers. Above each vertebra that is in optimal condition, the pendulum will be in its search position. It may move clockwise. But when the pendulum is suspended over an injured or displaced vertebra, it will move anti-clockwise or jump about irritably. The quicker and broader the diameter of its movement, the more severe the imbalance in the condition of the vertebra. Write down the vertebrae over which the pendulum indicated an imbalance. Now you can consult with the pendulum about each vertebra as to whether there is need for urgent chiropractic treatment, whether other treatment will help the condition of the disk, or if resting will suffice to restore balance.

Discovering allergies and food intolerance

One of the most oppressive and bothersome problems that people encounter during their everyday life is allergies of some type or intolerance to foods. The phenomena of allergies are many and varied, and sometimes the symptoms are so serious that the person has difficulty functioning normally. Finding the reason for the allergy is a long drawn-out process and not always pleasant, and sometimes even at the end of it the doctors do not find a clear answer as to the cause of the allergy - the allergen.

Diagnosing allergies or food intolerances

The diagnosis is greatly dependent on the allergy's degree of development. There are three stages:

The acute stage - Upon each exposure to the allergen or problematic food, there is likely to be an acute reaction that is immediate and obvious, and if one pays attention to this stage, it is very easy to take care of it.

The adaptation stage - The body adapts itself to the antigen and its response becomes delayed. Many times during the adaptation stage the body develops a craving for that food ("I can't live without it"), and then it is more difficult to identify the antigen. An elimination diet can help:

One takes the list of food items and eliminates a given food, which becomes completely forbidden to eat. We should suspect those foods that a person eats regularly and foods which have an allergic potential.

The degenerative stage - At this stage the physiological functioning is permanently disrupted and the adaptation has failed, and it usually isn't even considered that the person could be suffering from an allergy, because this is the stage at which the person is suffering from a disease of some sort. At this stage one cannot rely on the body's reactions to food in the normal way (as in the elimination diet), but rather there should be an extended period of non-exposure to the food, strengthening of the body, and then gradual exposure to the food.

The prevention and treatment of allergies includes a varied menu that never contains the allergenic factors or the causes of the intolerance. One can refrain from eating milk products or wheat products (allergenic items) for a day or two. Avoiding a suspect food for a period of four months can allow a return to the menu with no special problems as long as the food item isn't consumed very often. There must not be everyday contact with the allergenic food.

As many healers, as well as allergy sufferers themselves, know, the discovery of the allergen is the first and most important stage in the prevention and treatment of allergies. This process can be long and tedious, and sometimes even at the end of it, the real reason for the food allergy or intolerance is not discovered. The allergy sufferers resort to difficult and expensive laboratory tests that often include numerous injections, something that makes the diagnostics difficult when dealing with children. But many adults also suffer from tests such as the Rast test (an expensive technique that is meant to measure the level of antibodies against specific foods. The diagnosis works only when there is an immune mechanism), the FICA (a test that is very similar to the Rast but more expensive, measuring the entire antibody system in the blood); or the cytotoxic test in which they take blood with white cells, mix them with antigens of certain foods, and see how they handle them under the microscope. If they see that there is cell damage, it means that there is an allergy. The test is not always accurate, and depends on the skill of the technician. The more common test, and a less pleasant one, is the Patch test - an injection of substances in which the person is forced to undergo numerous jabs with needles, and which is not so reliable because when in a different state of health, the person will not get the allergy. It is also uncomfortable. It is meant to identify unequivocal conditions such as an allergy to the dust mite, olives, grass, or sand.

The non-laboratory diagnoses are considered to be easier and sometimes even give quicker results. Amongst these tests are electrical acupuncture test (a test in which the patient holds one electrode in his hand and they place food on an aluminium tray and go over all sorts of selected points); the kinesiological test, a method that examines muscular resistance (the food is usually placed on the patient's body or in his hand and the resistance of his muscles is checked); or a pulse test (the pulse is checked 15-20 minutes before and after eating the suspect food (absorption time). A rise of 10 or more heartbeats usually constitutes an indication of intolerance or allergy.

Checking food allergy or intolerance using the pendulum

Checking allergy or intolerance to food using the pendulum works in a similar manner as the non-laboratory methods, but is simpler, and can be performed by either the therapist or the patient. The principle by which this method works maintains, similar to the kinesiological test and the pulse test, that when there is a substance near a person to which his body objects, this will express itself in resistance of the body's various systems to that substance. In principle, the body knows exactly what things are good for it and which ones cause it to have undesirable reactions.

If you, one of your family, or one of your patients suffers from one or more of the many symptoms described above, and no medical reason has been found for the presence of those symptoms, the first stage will be to check whether the source of those symptoms is an allergy. To illustrate this, let's suppose the patient suffers from migraines.

Sit in a room free of any disturbing energies and take deep, slow, comfortable breaths. Declare your objective and ask the opening questions. If the patient is not present with you, it is desirable to place a picture of him on the table across from you or to write his name (his mother's name can be added) on a sheet of paper in front of you.

Now hold the pendulum and ask: "Is the source of the migraine that [patient's name] suffers from an allergy?" Watch the pendulum's movements. If a smooth, clear, affirmative movement occurs, note that the source of his problem lies in allergies. If a negative movement occurs, you will have to search for the reason for the symptoms in a different problem.

If the pendulum jumps around erratically, does not move, or its movement is not smooth or is ambiguous, you will have to clarify the question further. Ask: "Is there an allergic reaction involved in the migraine the patient is suffering?" If the answer is affirmative, it indicates that the disease involves several causes, including emotional causes and perhaps even spiritual ones. In such a situation, it is worth trying to check the body's aura, which appears in the chapter Healing Using the Pendulum, in order to pinpoint the additional levels affecting the origin of the migraine.

After having found out that the source of the illness - a migraine in this example - lies in allergy, or an allergic reaction is involved in its occurrence, ask the following question: "Is the allergy to a certain food?" Because this is one of the most common reasons for an allergy appearing, it is likely that the answer will be affirmative. If you receive a negative answer to this question, you will have to check additional factors which are not food-related. The most common of these is pollen, dust, the house-dust mite, pine, animal hairs, bouquet and flower fragrances, cleansers, soap, and the like. You must take into account the patient's living environment; for instance, if he works in a production plant of any sort, ask him about the substances with which he comes into contact - directly or indirectly - for example, sawdust, various chemical agents, and other such things. You can sharpen your questions more and more, according to your medical knowledge.

Diagnosing the food that causes the allergy or intolerance

After finding out that the patient suffers from a food allergy, the next stage will be to discover which food(s) the person is allergic to. This check requires attention and extensive investigation. Have the patient write down on a piece of paper all the types of food that he eats regularly. Now arrange them according to groups, in any manner that you like, as long as they are divided into various categories. An example could be: carbohydrates, proteins, fats, etc.; or fruits, vegetables, grains, etc. Write down the foods arranged into categories on a sheet of paper.

Now there are two methods at your disposal. If you are doing the check for self diagnosis, the first method is more highly recommended. In general, the first method tends to work more quickly and with greater accuracy, although the second method has also been proven to be efficacious.

First method: Checking the food itself

Using the pendulum, start out with a methodical check of the foods when they are laid out in front of you. Hold the pendulum in your hand with one of the foods placed on the table before you - a piece or a loaf of bread, for example. Hold the pendulum above the food being checked while asking: "Does this food cause me [or the patient use his name] allergies?" Or: "Do I [or the patient] have an intolerance for this food?" If the pendulum produces an affirmative movement, then this food comprises an allergen that activates the allergic response mechanism in you or the patient or the family member.

Continue checking each of the rest of the foods. Many people tend to be allergic to more than one food.

Second method: Checking the food by making a list

In this method, a list of food organized into comprehensible categories replaces the checking of the food itself. Write down on a sheet of paper those foods which the person eats on a regular basis. Divide them into groups according to each criterion that was chosen (for example, carbohydrates, proteins, etc. or vegetables, fruits, milk products, and so on). Record each group on a separate piece of paper. At the top of the sheet write the person's name (various therapists also write the name of the patient's mother). The list should be according to the following formulation - if, for example, you chose to divide the foods according to the major food groups.

For this example we will use the name Jake, son of Rachel. Under the heading "Carbohydrates," we included the foods which contain a high percentage of carbohydrates.

Jake, son of Rachel
Carbohydrates

Whole wheat bread	White sugar
White rice	Honey
Cornflakes	Barley
Yeast cakes	Doughnuts
Cheese cakes	Noodles
Sweets	Chocolate
Puffed wheat	

Now pass the pendulum over each food in the list while asking: "Is rice the cause of the patient's allergies?" If the pendulum gives a weak and unclear affirmative answer, phrase the question in a more precise manner: "Is bread involved in the patient's allergic mechanism?" (because it is likely that there are a number of components in the allergic mechanism, some of which are dependent on the patient's mental and physical condition at a given time, and bread does not always constitute a trigger for the development of an allergy, except in certain circumstance that involve different mechanisms). Mark the foods that the pendulum has pointed out as causes of allergy.

If you have completed the check of the foods that the patient eats on a daily basis and have still not received an affirmative answer regarding the kind of food that causes the allergy, you will have to make up a list that includes foods that the patient eats on a less regular basis, but could well be the causes of the allergy.

If at the end of the checks, you have received a great number of allergenic foods, you will have to check whether they contain a common ingredient that is probably the cause of the allergy. For example, if you have discovered that applesauce from a can, ketchup, and canned tuna all cause the patient to have an allergic reaction, it is likely that the preservatives present in all of them are in fact the cause of the allergy. Check this out using the pendulum. There are quite a few cases known in which the pendulum indicated almost all the fruits and vegetables placed on the table or appearing in the list. In this case, it is probable that the patient is not allergic to fruits or vegetables themselves, but to the pesticides that do not completely rinse off even after careful washing. In such an instance, the patient will have to switch to eating only organic fruits and vegetables.

9

The pendulum and our spiritual development

Choosing the right book

One of the most important uses made of the pendulum, particularly among those who engage in the fields of metaphysics, healing, Kabbalah, clairvoyance, and other New Age fields, is the correct selection of books about using the pendulum. At present, as we enter the New Age in which awareness and spirituality are slowly becoming the preserve of more and more people, the number of books dealing in the various fields of spiritual development, expanding consciousness, and metaphysics is increasing at an enormous rate. The problem is that not all these books in fact come from the same infinite source of truth and enlightenment, and many of them contain information that is confused, erroneous, and misleading. Every person who is nurturing his spiritual development must know how to choose only the books that contain true knowledge that is free of biases and special interests, and that is uniquely appropriate to his present spiritual development. In order to prevent taking in information that is superfluous and misleading, you should use the pendulum when you are about to purchase or borrow a book. The technique is very simple - all you have to do is hold the pendulum above the book you want to ask about and ask it: "Is this book appropriate for my spiritual development?" Or: "Is reading this book right for me?" This way you can also know which of two or more books dealing in the same subject matter will be the most suitable and right for you to purchase.

Choosing a teacher

When you are about to choose a teacher or instructor in any profession, and all the more so in the metaphysical fields, the compatibility between you and the teacher or the group with whom you plan to work and develop is of exceeding importance. Many teachers in areas of the occult do not have a specific kind of training, and that is because it is difficult to know what the required training in such areas is. There are wonderful teachers who received their gift for teaching from Superior Powers, and there are teachers who through their own spiritual development learned how to transmit information and encourage growth and development in other people.

On the other hand, as in every area, there are also charlatans. Many people join various groups that deal with various areas of metaphysics and awareness, ranging from yoga to Kabbalah. Sometimes a charlatan teacher, who is not genuinely aware or who has interests other than realizing his personal destiny and helping other people with their development, can cause a person to be drawn away from the true path he needs to take. Sometimes the choosing of a teacher is done on the basis of friends' recommendations, advertisements, or testimonials. This is not always the right teacher who is appropriate for you at this stage of your development.

When you are about to decide on a teacher or a group, you would do well to employ the pendulum's help. The simplest method is to write down the names of the teachers or groups on a piece of paper and hold the pendulum above each name while asking: "Is this the teacher [or group] that is right for me?" If the pendulum gives an affirmative answer for more than one teacher or one group, ask: "Is this teacher [or group] the best one for me?" You will probably be surprised to find that teachers who are well known and reputable in their professions can elicit a definitive negative from the pendulum, while other teachers who are less renowned and familiar will elicit an affirmative response. This does not necessarily mean that these teachers are charlatans or unsuitable for teaching the profession, but rather that the teacher who was chosen probably has more to give you at this period in your life, or that the personal compatibility and chemistry that is likely to take place between the two of you at this time is greater and more meaningful. This technique is, of course, also excellent for choosing a school or college.

The pendulum and meditation

Meditation is a state in which the brain rests from any thoughts or emotions, and concentrates on one idea or a particular object. Various meditational stages bring the person's brain into a state of alpha waves, the condition experienced moments before falling asleep, in which the brain functions at a different speed and with a higher sensitivity for taking things in than during regular waking states. Meditation is known for contributing to a healthy and serene life, releasing tension, alleviating negative emotions, and increasing creativity and the ability to focus. In addition, the state of relaxation that has a positive effect on the nervous system also affects muscle tone and causes muscles to relax, which strengthens the body and enables its organs to relax and renew themselves. Meditation is one of the most outstanding tools for spiritual development and growth.

There are numerous methods of meditation. Basically they all include conscious breathing (focusing on the breath and various breathing techniques), relaxation of the body, and mental calming. There are methods that concentrate on using a certain mantra, on focusing on parts of the body or chakras, dynamic meditation - movement, sound meditations, hypnotic meditations, and an enormous variety of additional forms of meditation.

Because of the differences between people, the form of meditation that is suitable for one might be totally inappropriate for another.

Apart from the investment of time and effort in reaching a meditative state that is likely to contribute very little to the person, there are types of meditation that may open certain doors in the person's consciousness that should not be opened, or take up a great deal of his time attempting to achieve a meditational state that is impossible for him to achieve at the given time. Because of this, it is important to use the pendulum when you are interested in choosing the form of meditation that is appropriate for you. Furthermore, the pendulum should be used for determining the length of time to meditate, and the place in the home in which is most appropriate to meditate. The place is of great importance because during meditation, the person's sensitivity to absorbing things is highest. If a person meditates in a place in which base or negative energies of any sort were present previously, those energies are likely to penetrate into his electromagnetic field, especially if he is not an experienced and practiced meditator. The length of time of meditation is also very important. When a beginner meditator meditates beyond the limits of his ability, it is similar to a beginner athlete who runs too far and exhausts his muscles. Use the following method to determine the suitable meditation time for you:

Draw a face of a clock divided into sixty minutes; this can be done in groups of five. After setting forth the objective of the check for yourself, and asked the basic questions that must be asked before using the pendulum ("Can I ask this question? Am I allowed? Am I ready?"), and having emptied your mind of all emotions and thoughts, hold the pendulum above the drawing of the clock, over the line of the first five minutes, and ask: "Am I capable of meditating for five minutes?" If the answer is affirmative, you can continue and as: "Am I capable of meditating for ten minutes?" and so forth. When the pendulum performs a negative movement, you have reached the limit of how long you can meditate.

There is a wonderful mutual relationship between the use of the pendulum and meditation. In many aspects, using the pendulum resembles meditation, as it requires concentration, relaxation of the body, and detaching oneself from feelings and emotions. Therefore, meditation will help you develop your skills in using the pendulum, while regular use of the pendulum will help you get into the meditational state more easily.

10

The pendulum and astrology

First stage: Determining the person's zodiac sign

One of the best-known problems in astrology is information concerning the **precise** time of birth. The pendulum enables us to find the missing data very accurately, and to prepare a precise and reliable astrological map.

Place a sheet of paper in front of you on which the twelve signs of the zodiac are listed in order. The symbol of each sign should appear next to it. Above the list of the signs and their symbols, write the person's first name. To give an example, we will use the name Brian.

Brian

Aries	Libra
Taurus	Scorpio
Gemini	Sagittarius
Cancer	Capricorn
Leo	Aquarius
Virgo	Pisces

After having declared your objective and having asked the opening questions, hold the pendulum and pass it over each sign in turn, asking: "Is this Brian's sign?" Or: "Is Brian's sign Taurus?" and so on, until the pendulum makes an affirmative movement over the person's sign.

Second stage: Finding out the person's ascending sign

In order to discover the person's ascendant, repeat the first stage in exactly the same manner, but the question you now must ask is: "Was Brian born under Capricorn in ascent?" Or: "Was Brian born under Taurus in ascent?" and so on, until you receive an affirmative answer from the pendulum.

Now all that is left for you to do is to discover the precise angle of the sun's rising in the sign.

Third stage: Finding the sun's angle in the sign

In order to find out the angle of the sun, write down the person's name on a sheet of paper, and underneath it, write the numbers one through thirty, as follows:

Brian

1	6	11	16	21	26
2	7	12	17	22	27
3	8	13	18	23	28
4	9	14	19	24	29
5	10	15	20	25	30

Grasp the pendulum and pass it over each number, while focusing strongly on each one, and think of it in a conscious manner. Ask: "Is this the rising angle?" When you reach the correct angle, the pendulum will gyrate in an affirmative movement.

These data will help you find out the zodiac sign, the ascendant, and the angle of the rising sun. By doing this, you will discover many of your [the person's] character traits, hints and explanations, including negative personality you will understand that certain traits that may have been perceived as distressing personal faults in fact derive from the sign under which you [the person] were born. It is easier to identify and correct these faults when you are not doing so with a guilty conscience, and are aware of the fact that in spite of the position of the stars in the sky being one of their causes, you have the full power to change those characteristics if you only put your mind into it.

The astrological map (horoscope), for all its many advantages, suffers from one prominent drawback: it is capable of giving a person reliable information about himself and of serving as an excellent diagnostic tool, but here, in essence, its role is likely to end. It tells the person about his traits and habits, but does not give him the necessary tools to make the desired changes. And thus, for hundreds of years, the horoscope served solely as a diagnostic tool, without the astrologists being able to develop it into something more than that and transform it into a therapeutic tool as well. Now that astrologists have begun using the pendulum as a supplementary tool in drawing up a horoscope, this science has begun to progress to the point that in the future it can become a combination of diagnosis and treatment.

As regards the planets and their characteristics in depth, we will discover that each planet has very clear characteristics which are seen in the nature of the person who is born under its sign. Each planet has its own principles and an energetic path drawn from it that affects the person's soul. Our physical and emotional health, as a microcosm, depends on harmonious composition and balanced cooperation between the components of this energy. When this balance is upset, the person is likely to suffer a lack of physical or mental balance. Already in ancient times, each planet was attributed an influence on an organ or system in the human body, on their functions, and on mental functions. If we accept the premise that man is a reflection of the universe, it is easy for us to understand that just as an apparent lack of harmony between the planets of the heavens and their orbits have an effect on the universe itself - these conditions of disharmony are likely to manifest themselves as a lack of balance between elements and "organs" of the universe, for example, by causing earthquakes, floods, storms, fires, and the like (large floods, for example, are like the element of water conquering the element of earth, and so on) - so a lack of balance between the energy of the planets in the body and soul of the human being creates a clash between different elements, and within the functioning of organs of the body itself. As the sages and ancient philosophers, who were extremely experienced in astrology, put it: "As it is above, so it is below." They went even further and determined that if a person concentrates on learning about himself as a microcosm of the universe, he will also arrive at a complete understanding of the wonder of the universe.

11

Using the pendulum for healing

Detecting blockages

As we can see with our eyes in this seemingly "physical" world, lines of flow or progression are likely to have certain obstacles along their route. A person walking along a path and encountering a large rock will look for a way around it, something that doubtlessly causes him to deviate from his original path. It is the same with a strong current of water that divides and goes around the various obstacles in its course. This deviation of the water from its course because of the obstacle in its path creates a sort of small island around which the water flows, while the speed of its current diminishes. In the same manner, water flowing along a pipe in which there is a blockage in a particular spot will cause the pipe to swell. If the water is not turned off, the pipe will burst. The same thing happens with energy.

One of the primary goals of healing is the removal of blockages from the human body. This energy, called "chi," the life force, or electromagnetic energy, flows in the personal energetic network of every person, similar to the flow of blood through the circulatory system. When a blockage occurs in the circulatory system, a health problem is created. The solution is the release of the blockage, whether through diet, physical activity, or in more serious cases, bypass surgery. It is the same when a blockage occurs in a person's energetic system. Energy blockages express themselves in all areas of life, ranging from one's physical, mental, and spiritual health, becoming accident prone, and having bad luck, to finding unsuitable romantic partners or jobs, or friends who are undermining. In other words, when a spiritual blockage affects the person's electromagnetic field, its action will manifest itself on all levels. Healing is one of the best ways to release these blockages. Many people are excited to find out that after several healing treatments - sometimes even after just one treatment - their lives change immeasurably. One of the reasons for this is the release of those blockages, the result of which are seen in everyday life, in physical and mental states, in relationships, in their careers, and in a generally much improved feeling.

Using the pendulum to locate energetic blockages

Before using the pendulum in healing, go over the basic rules discussed in the chapter, "Rules for using the pendulum." Cleansing the pendulum before working with it is of cardinal importance, especially between one treatment and the next. The human body radiates energy from within, and these energies must not be radiated into the patient's body. To do this, we need to cleanse the pendulum by holding it under running water, placing it on a quartz crystal cluster, and if the healer is experienced in activating power of thought, he can cleanse it by means of that as well. An important rule: the healer must remember that he is serving a tool of the cosmic energy for purposes of inspiring it on earth. He must remember that he is only a conduit, and the energies passing through him are supreme cosmic energies and not his own personal energies. He does not "give" or "receive" energies, but rather serves as a channel for divine energy. It is vitally important that he empty himself of all emotions and thoughts before the treatment, and protects his own aura. An acceptable protection, which involves infusing oneself with energy, is the image of a golden ray of light descending from above, entering through the crown of the healer and descending along his spine, leaving his body via his coccyx, continuing downward and connecting with the earth. From within the earth the golden light ascends upward, and envelops the healer in an ellipse of golden light. The healer inhales from this ellipse into himself and exhales it out to the ellipse. He should request love and guidance. This technique of infusing energy and protection can also be applied to the patient (with the healer seeing him in his mind's eye surrounded by an oval of golden light). In no case must the healer and the patient be surrounded with the same ellipse of light.

Requesting permission to treat

An important stage before beginning the healing itself is asking the patient's permission. Even if you are doing healing from afar, you should ask the patient's permission to treat him. If you are unsure as to whether he'll understand the process of healing, you can request his permission to send healing energies, or to pray for him, or explain in any other manner that the patient will understand. Just as we would not enter a strange house without being invited in, it is also best that we not "interfere" with another person's energies without asking his permission beforehand. If the patient is too weak or ill to answer, ask permission from one of his relatives. In a case where you don't have any possibility of making contact with the person whom you are interested in treating or checking from afar, ask permission from the forces of the universe for your actions, and request their assistance. Observe the rules for tuning the pendulum that are described in the chapter, "Rules for using the pendulum," and pay particular attention to the question: "Am I allowed to treat this person?" and the pendulum's response.

Locating blockages and holes in the aura

Before beginning the process of healing and the examination, give yourself a little bit of time to connect with the universal force, by whatever name you call it, to create protection, emptying and filling yourself, and make sure to ask the four questions: "I am interested in checking for blockages and holes in the patient's body. Can I do this? Am I allowed to do this? Am I ready for doing this?"

Now, after having received affirmative responses to your questions, ask the patient to lie down on his back on the floor or on a mattress or blanket. There are some healers who say that during healing the left hand should always be on the upper part of the patient's body. To do this, sit at the patient's left side. Many healers help the patient loosen up and relax by using the "rocking" technique: Place your right hand in the region of the patient's coccyx and your left hand on the seventh cervical vertebra (the first vertebra that sticks out between the nape of the neck and the rest of the spine), and begin rocking the patient with light, gentle movements. Ask him to breathe slow, deep, comfortable breaths. Make sure that throughout the treatment your own breathing remains deep, comfortable, and conscious. If the patient is not relaxed enough, ask him to concentrate on his breathing. This will help him relax his brain and his body. Concentrate, and ask the force of the universe to help you become a channel for healing.

Now hold the pendulum in one hand and put your other hand at the top of the patient's spine, on the seventh cervical vertebra. Pass your fingers slowly down the vertebrae, one by one, along the patient's entire spine down to the coccyx. Pay attention to the pendulum's movements. As long as the condition along the spine is positive, the pendulum will gyrate in the search position movement. The moment you locate a spot in which the balance is upset, the pendulum will move from its search position to gyrating in some sort of movement. It doesn't matter what sign the pendulum produces; the moment it alters its gyrations from the search position movement, it is a sign that there is a hole or a blockage in this area of the spine. The pendulum will continue moving until you reach the end of the hole or blockage, and further along the spine it will resume its regular search position until it reaches another blockage or hole. Note to yourself where the holes or blockages are. Stop when you reach the coccyx.

The procedure for balancing the chakras using the pendulum

If you are experienced in the use of crystals, it is best that you use a crystal pendulum for this purpose. White quartz is excellent for this purpose, but you must cleanse the pendulum extremely well before and after its use.

After you have declared your objective and checked your ability to carry out the task, have received permission to do it, and are ready for it, empty yourself of any emotions and thoughts, and fill yourself up with the universal energy, create protection, and request assistance and power from the universe. Ask the patient to lie on his back on a flat surface such as the floor, a mattress, or the like.

Hold the pendulum above the first chakra, the base chakra. A woman's base chakra usually contains yin -

negative - polarity, while a man's base chakra contains yang - positive - polarity. When you hold the pendulum over the base chakra of a male, the pendulum will move from its search position to a positive gyration, and will continue moving in this direction for a while. When the pendulum returns to its search position, this means that the chakra is balanced (the balance is not always complete, as each person has the ability to take in a certain amount of energy and sometimes you will need to repeat the procedure another time; in any case, the return to the search position indicates that you need to move on to the next chakra). Move upwards slowly, until the pendulum is above the patient's sex chakra. Now the transition of the pendulum from the search position will be to a movement in the opposite direction to its movement above the first chakra (in a male, a negative movement - yin, and in a woman, a positive movement - yang). When the chakra has achieved the maximal balance it can at the moment, the pendulum will return to its search position. Continue on to the next chakra, the third. Remember that the polarity alternates between one chakra and the next, according to the person's gender. If the pendulum remains in its search position over a given chakra, skip it and move on to the next chakra. For some reason, that chakra should not be balanced right now.

After you have finished balancing the chakras, thank the universal force for allowing you to be a channel for its light, and thank the patient.

Detecting blockages without the person present in the room:

Detecting the blockages must be done in a suitable room - clean and well-ventilated and clear of any undesirable energies. After selecting an appropriate room, and using a suitable and cleansed pendulum, the healer should make a schematic sketch of the human body and at the top of it write down the name of the person and the name of the person's mother. To illustrate, we will use the name of David son of Marge.

The healer places the diagram on a clean, empty table or other surface, and scatters samples taken from the body of the person being treated. These samples can include tiny portions of hair, skin, fingernails, saliva, blood, etc. If the healer has good visual or audial abilities, he can visualize the image of the patient before him for a few minutes, or "hear" his voice. If he succeeds in tuning in to the patient in such a manner, this method can replace the samples. Now, without asking any questions, the healer passes the pendulum over the various parts of the body detailed in the schematic picture of the human body. It is usually customary to first go over the limbs, and afterward pass the pendulum along the line of the chakras (base, sex, solar plexus, heart, throat, third eye, and crown - above the crown of the head). At this stage the healer will observe the movements of the pendulum and will discern where it gyrates in a positive movement and where it gyrates in a negative motion (according to the way he has programmed it). With this technique, it is easier to use the pendulum programmed to positive and negative movements by circling clockwise or anti-clockwise, with movement in straight lines signifying energy that is truncated, in which case one should stop, refocus, and try again. Experienced healers recommend passing the pendulum over the entire body three times, and note for themselves the region(s) in which the pendulum gyrates in a negative movement. Those are the areas on the patient's body that contain energetic blockages.

After the healer has located the blocked areas, he needs to draw up a list of all the body parts or organs found in those areas. For example, if there is a blockage in the area of the throat chakra, the healer must note down the thyroid gland, the vocal cords, the trachea, the esophagus, the parathyroid glands, etc. To do this, he needs as thorough a knowledge as possible, in addition to knowledge about the connections between the chakras, the meridians, and the different parts of the body. For this, it is recommended to use an anatomic atlas or the "Yellow Emperor's Book of Internal Medicine."

For purposes of illustration, let's assume that the blockages are located in the chest area. The healer must write up a list that includes the heart, lungs, breasts, intercostal muscles, etc., and of course also the bones, and the major blood vessels and endocrine glands.

The list should be placed above the diagram of the human body. Now the healer should pick up the pendulum and pass it over each of the names while asking: "Is the energetic blockage located in this organ/part?" A negative response by the pendulum indicates that the organ is healthy, and an affirmative answer tells us that there is a blockage and a problem in this part. The healer will usually find that several organs are blocked.

Now, in order to obtain more specific information about the type of problem, the healer concentrates on a given organ, requests guidance, and, while holding the pendulum, asks: "Does this organ suffer from an excess? Does this organ suffer from a deficiency?" and so on. If the healer has medical knowledge or has books about organ function, he can receive more precise answers about the nature of the problem. For instance, if he has identified a problem in the thyroid gland, he can ask: "Is the gland overactive? Is the gland underactive? Does the problem derive from radiation?" and so on. In addition, he goes back and checks the systems associated with the affected organ. If the problem is in the thyroid gland, for example, he will again pass the pendulum over the patient's head to see if there is negative vibration in the region of the pituitary gland, or will write down the name of the endocrine gland on paper and ask; "Is the source of the problem located in the pituitary gland?"

    The advantage to this method is that the patient's presence in the room is not necessary and it can be carried out from any distance, of course. If the healer is optimally focused, in touch with his intuition, and has medical knowledge, his diagnosis will resemble a medical diagnosis in every sense, without the discomfort and the many complications implicit in a diagnosis of that type. It should be remembered that in certain cases, a medical diagnosis cannot be replaced by a diagnosis using the pendulum. The healer must be aware that if he discovers a particular problem that also requires a medical diagnosis, he must refer the patient for an appropriate medical checkup that will confirm his assumption. In addition, the healer must be extremely sensitive and know how to break the news gently to the patient. If the healer is blessed with a good ability for visualizing and connecting, he can now move on to an additional stage of treatment: he can visualize the patient lying before him or sitting in front of him and, using the palms of his hands, transmit energy to the patient while concentrating on the problematic areas; he can make the movements of extracting and infusing energy, cleansing, and energizing, as is appropriate for the problem.

An additional advantage of this method is that the very vibrations of the pendulum itself while they are passing over the diagram, especially if the healer is tuned in to the patient well, help to open up the energetic blockages, to move energy along, and to improve the aura sheath.

Exactly the same method can be used in the presence of the patient, with the patient lying relaxed on a bed. Observing the rules of healing noted above, the healer passes the pendulum over the patient's body just as he did over the diagram, and asks the questions. It is best to cleanse the patient's aura before the treatment and to ask him to concentrate on his breathing so that he won't give off energies of excitement, anxiety, and so on, which could interfere with the diagnostic process. This method can also be applied in the presence of the patient while using a diagram, and then there is no need for samples. Each healer should choose the method that suits him, with pure motivation and devotion to his work.

Locating the blockages in the different bodies

In order to receive more detailed information about the essence of the patient's problem, or as a technique in its own right (particularly useful when the patient seems to have no physical symptoms and most of his problems are concentrated in the mental or spiritual realm, and have not yet shown up physically), the therapist can use the technique of locating blockages and imbalances in the various bodies.

In this technique, the four main bodies will be checked: the physical body, the ethereal body, the emotional (astral) body, and the mental (thought aura) body. Some people improve on the method and also include additional bodies, but the method is presented more simply here:

The therapist takes a sheet of paper and writes the name of the patient at the top. If he is using this method as a continuation of a previous diagnosis, he also writes the name of the organ in which a problem was found under the patient's name. If he located a problem in more than one organ, he must prepare an additional diagram for each organ.

Now the therapist should write the names of the bodies under the heading:

> The physical body
> The etheric body
> The astral body
> The mental body

After the therapist has cleansed himself of all thoughts and emotions, declared the objective, and asked the opening questions, he points with his left hand (or with his right hand if he's left-handed) to the physical body, and, while holding the pendulum in his other hand, asks: "Is the patient's physical body healthy?" For this question, an affirmative movement by the pendulum will indicate that the physical body is healthy, and a negative movement will indicate that there is a problem in the physical body.

If the therapist is using this technique as a continuation of the first technique and has written the name of the affected organ under the patient's name, he will ask: "Is the essence of the problem in the physical body?" There are therapists who do not ask any questions, but rather pass the pendulum over the words "physical body" while observing the movements of the pendulum. If the pendulum gyrates in a positive movement, then the blockage is located at the physical level. Attention must also be paid to the vibrations of the pendulum - stronger, more rapid, weak, having a wider or a narrower diameter. This observation is important because, as mentioned, the problem can occur in several of the bodies or in all of them, and hence there is importance to the force with which the pendulum indicates the existence of the problem in the various bodies.

Now the therapist will repeat the check on the other three bodies. We must remember that a problem in a given organ in the physical body can show up on all three other levels as well. If the pendulum indicates a problem in one body only, the patient's state of health is relatively good. A problem in two bodies makes it clear that some improvement is needed. A problem in three bodies indicates a more basic or ongoing problem that has already permeated more body levels. First of all, a lot of rest must be recommended to the patient in order to give him respite so he can rehabilitate his auras and the life force flowing within him, as well as avoid exacerbating the problem and negativity of the bodies. A condition in which all four of the bodies are suffering from negativity - deficiency or excess or blockages - is an emergency situation. If we repeat the checkup again several times (preferably over time-periods of several days or more) and again discover such results, it is a sign that the patient is about to collapse. In general, the more bodies involved in a problem, the more serious the patient's situation is.

If we have checked a given organ (that is to say, we have written down the name of the problematic organ at the top of the list underneath the patient's name) and we know through having used the first technique that there is a problem in that organ, and we nevertheless received a different answer (that is, that there is not a problem in the organ) and we didn't find blockages in this checkup, there are several possible reasons:

* There is the possibility that there was a mistake in the examination. To find this out, it should be repeated again on the same day or the following one.

* The healer should check his degree of concentration and alertness. It is worthwhile for him to rest and relax before carrying out a repeat examination.

* It is possible that the process of the diagnosis itself using the pendulum (especially if the healer used a quartz or crystal suited to the patient and his problem) released the energy blockage.

* It is possible that there is strong radiation from a blockage in a different organ that is affecting the organ we checked, for example, radiation from the pituitary gland onto the thyroid, or the influence of a neighboring organ such as the liver on the gall bladder. Therefore we need to continue checking the rest of the organs that are associated with a problem, and to come to a final conclusion only after an comprehensive examination.

After we have received clear answers regarding the essence of the problem and we know which organs have blockages and in which bodies the blockages are found, we can treat the patient according to the bodies in which we located the problem.

In the physical body, we will use physical means such as diet, physical activity, massage, Shiatsu, reflexology, healing herbs, and medical treatment, as necessary.

In the ethereal body, we will treat using means such as movement, dynamic meditation, conscious breathing, aromatic baths, a lot of rest, sunlight, and fresh air.

In the emotional body, we will treat using art therapy, color therapy, Bach Flower Remedies, techniques for the release and letting out of non-positive emotions, techniques for relaxation and ensuring a supportive and enabling social environment.

In the mental body, in which the therapy is likely to be longer and more drawn out, the treatment will involve expansion of self-awareness. Sometimes problems in this body are created by overuse of thought and analysis, self-criticism and negative thinking. It is very important to locate the negative thought patterns the patient is using, which can be diagnosed according to the affected organ. (It is a good idea to seek the assistance of Louise Hay's excellent book, "You Can Heal Your Life" and the table that appears at the end of the book). In addition, relaxation exercises and meditation, and exercises to activate the right hemisphere of the brain, can help in this situation. If the condition of the mental body continues to be non-positive, the patient must cooperate and make a conscious effort to get rid of the negative thought patterns to which he is apparently prone.

Treatment and diagnosis using the pendulum allows us to treat a person (or ourselves!) as a whole and complete unit. Use of the pendulum for diagnostic purposes makes it significantly easier on the healer and increases the accuracy of his diagnosis.

Matching color to the aura

As we know, our auras comprise all the colors of the spectrum in different combinations. The astral body carries within it a very delicate balance of colors. As long as this balance is maintained, the person feels healthy and calm. When this balance is upset, different diseases begin to occur, as well as various types of discomfort, unbalanced moods and tiredness or general nervousness. When you find yourself in such a situation and do not succeed in discovering the cause of the mood swings or the reason for the sudden weakness and fatigue, it is possible that the balance between the colors in your aura has been upset. To find out, carry out a simple check. Write down a list of colors without writing your name beside them. Again, it is preferable to write the names of the color using pens of the matching color. After you have emptied yourself of thoughts and emotions, asked the opening questions, and taken a few deep, slow, comfortable breaths, hold the pendulum above each color and ask: "Do I lack green? Do I lack pink?" and so on. Mark the colors over which the pendulum indicated an affirmative sign and continue the check. You are likely to find that you have several colors lacking.

Occasionally, after an emotional storm, a fight, an argument, or a reaction to distressing news, you are likely to find out that you lack many colors, sometimes even all the colors of the spectrum.

After you have identified the color(s) you are lacking, you can rectify the condition by wearing clothing of those colors, lying or sitting in front of a lamp that gives off colored light (it is possible to even take a regular table lamp and paint its bulb in the missing color, or wrap it in cellophane of that color). Sit or lie in front of the lamp for several minutes, according to your feeling. If you are not sure what length of time is suitable, you can check this out by presenting the question to the pendulum.

A simpler way to receive the color you are lacking is to empty your mind of all thoughts and emotions, take a few deep, slow, comfortable breaths while sinking into a meditative state, close your eyes, and ask the Infinite Life energy to send you the required color. If you are extremely connected, the color will appear before your closed eyes, and you can visualize how it envelops and surrounds you, permeating your body, and filling it until it comes out of your body and surrounds you with an ellipse of illuminating colored light. Repeat this technique with each missing color, resting a bit between one color and the next. You are likely to find that several colors are sent to you simultaneously, each for a different length of time. Some people who have tried this technique when they were lacking all the colors of the spectrum, said that the color that was sent them was clean, pure white light. The reason for this is that the color white contains all the colors. Connecting with the Infinite Energy with confidence and faith ensures that the colors will be sent to you in the correct measure and frequency. If you have trouble applying this technique, you can ask a loving and aware friend to sit across from you, empty himself of all thought and emotion, request assistance from the Infinite Energy, concentrate on a piece of paper of the same color as the missing color or a piece of colored cellophane, and visualize how the color ray is being sent to you and enveloping you. If your request for assistance is sincere and he has the capability of connecting strongly with the Cosmic Energy, he may find that he does not need the piece of colored paper, and the ray of color appears before his eyes and is transmitted to you.

Another method is to meditate while you visualize the name of the color - the letters of the word - written in the color itself.

And so, as you have seen in this chapter, the possibilities for using the pendulum for purposes of healing and spiritual development are many and varied. After you have practiced the use of the pendulum for a while, try to incorporate it into other areas that are connected with spiritual development, while discovering for yourself ancient/new techniques. If you are a healer who receives patients, you will discover that the possibilities for using the pendulum are almost endless.

In conclusion, here is the story of Jill, a reflexologist who was helped by the pendulum for a slightly unusual goal, as it happens.

Jill, a young reflexologist, did her internship at a drug rehabilitation facility. She discovered that her work was benefiting the three patients whom she took on at the recommendation of the institution's director exceptionally well. She devoted herself to her volunteer work and found great satisfaction in it. The troubled lives of the recovering addicts and their sincere attempts to rehabilitate their lives touched her heart. She discerned that her treatments were helping them from both the point of view of health, since after the withdrawal, they were beginning to feel the terrible effects of the drug and were suffering from many difficult problems - and from the mental point of view. Within a short time there was a large gap between the physical and mental progress of her patients, and that of the rest of the patients in the institution.

Word about Jill and the success of her treatments soon

got around the facility, and many patients begged the director to refer them to Jill for treatment. But Jill, who was doing this work on a volunteer basis, as a young woman who was just starting to live independently, was working at two jobs. During the day she was a janitor in a kindergarten, and in the evenings she worked as a waitress. In spite of her great desire to treat as many recovering addicts as she could, she knew that her time framework and the necessity to earn a living would not permit her to give up any more hours of work. She decided to take on one more patient only.

When she asked the director of the facility to refer one additional patient to her, the director shared his indecision with her. Out of a list of eleven ex-addicts, he was able to eliminate six, but he was unable to decide between the five that were left. He explained to her that he had no alternative but to leave the decision to her, as she was the one who knew which of the patients would benefit most from the therapy. But in parallel, he made it clear to her that until she had chosen a patient, interviewed him, and signed that she would uphold medical confidentiality, he could not give her in-depth personal details about the patient.

Jill wasn't certain what to do. The information the director of the facility had given her about the patients didn't help her at all in choosing an individual patient. Everyone needed the treatment to the same degree. She shared her indecision with her teacher. The teacher, who had been engaged in broad areas of mysticism and holistic medicine for many years, invited her to her house, and asked her to write down the names of the five recovering addicts on a piece of paper. Now the teacher took out her pendulum and started passing it over the names. Over each name, she asked: "Is this person the most suitable one to receive reflexology treatments from Jill?" Three patients were immediately eliminated from the selection, as the pendulum made a clear "no" movement regarding them.

The teacher asked Jill to rewrite the names of the two remaining ex-addicts over which the pendulum had gyrated in an affirmative movement, and she again passed the pendulum over their names, repeating the question: "Is this patient the most suitable one to receive reflexology treatments from Jill?" Again both patients got a positive response, but the difference between the movement of the pendulum over the first name and its movement over the second was obvious. Above the first name, the movement was a clear, smooth clockwise circular movement, the diameter of the circle being particularly great, and the movement of the pendulum was relatively rapid. Above the name of the second patient, the movement was more hesitant, less smooth, and had a considerably smaller diameter. The teacher told Jill that the first patient was the more suitable for treatment. She explained that there are important criteria for treatment that no written information can give her, and the first of these is the compatibility between herself as a therapist and the patient. It was important that the patient be able to trust Jill, open up to her, and be free around her; and it was no less important that the patient have a genuine and sincere desire to rehabilitate himself, to persevere in the treatment. The importance of the energetic compatibility between them was great, and using the pendulum was the only method of finding this out in advance. Jill understood exactly what she was talking about, as a lot of time had elapsed before she had succeeded in creating trust with her first three patients, and more than once it happened that in their initial sessions they arrived late and occasionally failed to show up at all. She knew that a patient who was able to relax would reap the greatest benefit from the treatment, and the work with him would be easier and faster.

Jill informed the director of her decision. It became clear to her in an exceptional way within a few sessions that the choice was the correct one. This was a young patient, very keen to rehabilitate his life. After his first session, he already trusted Jill fully, and recounted his entire life story to her. He turned out to be the most persevering of all her patients, and the treatments were successful beyond her wildest dreams. The young patient, thrilled by the success of the treatment and the great relief it afforded him, both from the physical and mental points of view, wanted to know more and more about the various areas of holistic medicine. Jill recommended books on self-awareness to him, and these quickly became an integral part of his life, accelerating his rehabilitation, bringing him to an understanding of his past life, and helping him overcome the obstacles of the present. After a series of treatments, the patient decided to take a bold step toward rehabilitation, and took up studies in reflexology himself.

A few years later, Jill ran into him him at a reflexology conference. No one would have recognized in him the skinny, neglected, and confused young man that she had first met. No one would ever imagine that this person, now married with a family, had ever been trapped in the nightmare of drug addiction.

12

Using the pendulum throughout the house

In spite of its being such an important and expensive tool for significant discoveries in the physical, mental, and spiritual realms, there is nothing to stop you from using the pendulum as a tool in daily life. It can be employed constantly throughout the house in many different uses, from selecting the most appropriate wallpaper for the living room, to identifying a problem in a pet or houseplant, or to locating lost objects.

Some therapists, who use pendulums for locating and treating blockages, prefer to keep two or more pendulums. One of them is used in treatments, and the other is for more mundane purposes. But on the other hand, many prefer to use their personal pendulum whenever it is needed. Do as you personally think best.

Moving to a new house

All of us aspire to live in a place where the general energy is positive. This energy depends on numerous factors: the residential environment, which includes the energies of the area itself, and that of the neighbors and the people there, and the energy of the ground upon which the house is built - there are places in which the energy of the ground is not positive, and this has a serious impact on those living on it. This energy derives both from the type of the earth itself and from historical events, sometimes unknown, which took place here in the past. Of course, the energy of the house itself is most important.

When you are about to buy a new house or rent an apartment, use the pendulum for making this important decision. Employ it during all the stages of the decision and it should provide you with invaluable information.

The first stage is usually choosing the city or area where you are interested in purchasing, building, or renting a home. You will need to check whether indeed the energy of the selected place is harmonious and appropriate for you. This technique is simple, but requires concentration. Write the name of the place in which you're interested in living on a piece of paper. Declare your objective, ask the opening questions, and empty yourself of all opinions, thoughts, and emotions. Take several deep, slow, comfortable breaths. Put the piece of paper with the name of the area you are planning to live in written on it in your line of vision and concentrate on the name. A name, in general, is a form of connection and a vessel for the energies of the thing it is defining. The name of the place helps you to "tune yourself in" to its energies. Give yourself a little time to tune in to the place. Now hold the pendulum over the name of the place and ask: "How harmonious will the connection between my energy [or that of the person for whom you are carrying out the check] be with the energy of this place from the point of view of a good life, good health, and success?"

Observe the pendulum's movements. The stronger the positive movement is, the more harmonious the connection between you and the place will be. A negative movement indicates a lack of harmony.

If you engage in a profession or a particular field in which you attach great importance to the energy of the place (many artists see this as very important; therapists as well), ask: "How harmonious and beneficial will this place be for my creativity [or any other occupation]?" Observe the pendulum's movements.

After you have chosen the area in which you would like to live, you must choose the house that will serve as your residence. Take the pendulum along with you to the houses you go to see, and ask: "Is this house appropriate and harmonious for me?" If you receive an inconclusive answer while you are standing in one of the rooms, you should try to ask again in other rooms or outside the house itself. In any case, it is preferable to ask the question in several different places in the house, because it is possible that certain energies that exist in the house can affect the accuracy of the answers.

If you must select a certain plot of land for building a house, ask: "Is this piece of land harmonious and appropriate for building my house [or that of the person for whom you are carrying out the check]?"

After you have found a suitable house, you will need to check whether the energy present in the rooms is in fact suitable for you - clean and pure. This is of utmost importance. Many times people have lived in houses in which there were quarrels, energies of anger and sorrow were emitted, or other non-positive energies. These people found themselves affected by the energies present in the house, and experienced unpleasant feelings and changes in their mood or even in the course of their life itself, without understanding that they were being affected by the energy of the house. Sometimes a particular energy that is not non-positive, but is not suitable for the person moving into the house, can create situations of imbalance in him. These conditions can be improved, and the energies present in a house can be cleansed and balanced. But one must be aware of this. When you come to live in a new house which is likely to be suitable and correct for you in many aspects, make sure to check the energy in every place in the house. This check should be made when you are calm and comfortable, and it is best if you cleanse yourself of all opinions, emotions, and thoughts of your own in order to prevent your own emotions from influencing the vibrations of the pendulum. After you have declared your objective and asked the opening questions, stand in the room you wish to check, hold the pendulum, and ask: "Is the energy present in this room good for me?" If the response is negative, you will need to purify the room of the energies present in it. To do so, first clean the rooms of the house thoroughly. After the regular washing down, you can rinse the house with sea water or salt water, which have good purifying qualities. After you have cleaned the house and its surroundings well, use various tools for purifying the energy. Incense can be used - frankincense and sage are good for purification; it is desirable to go through the house with the incense, crisscrossing the rooms. Purifying aromatic oils (such as juniper, eucalyptus, frankincense, and so on) or purifying Bach Flower Essences (such as agrimony and crab apple - a few drops dripped into an oil burner with water, and a candle lit underneath) are also good. You can also use crystals or lighting "bakhour" (a mixture of Middle Eastern spices available at various markets, used to get rid of the "evil eye" and cleanse energies. The "bakhour" is placed on a frying pan or a large spoon and heated over the fire until it gives off steam. It is passed throughout the house. Places in which the smoke stops altogether sometimes require a more energetic cleansing).

You can also use the pendulum to choose the most

appropriate method of cleansing, the type of incense, the appropriate essential oil, or the crystals.

These methods of purification are also extremely important even after you have moved to the house in which you will spend your life. It is good to check the energies of the various rooms with the pendulum occasionally and to cleanse them as need be. This check is important, especially after people carrying a load of negative energy, or "energy drainers," have visited you home, or after arguments, disagreements, or other situations that cause an upset of balance in the home. Feng Shui spirals and Feng Shui methods can also be used to create harmony in the home. The following story will illustrate the importance of the selection of the residential area and the house itself.

Vivian, a young medical student, had been living in rented apartments from a young age. She had been "burned" many times. After having lived in three different places and having encountered landlords who were amazingly nice at first, only to reveal themselves to be inconsiderate and downright unpleasant later on, and after a lot of run-ins with the neighbors, mainly because of her large dog with which she would not part for anything in the world, she thought that bad luck was pursuing her in her choice of homes. By nature, she was a very energetic and friendly girl, who got on wonderfully with people.

In her last rented apartment, things came to a head. Two months after she had signed the lease, serious problems arose in the plumbing, electrical system, and gas supply. The toilet and sinks were permanently blocked, and the landlord, who had seemed to be nice and very understanding when she rented the apartment from him, turned out to be selfish and indifferent. He made it clear to her that he was not in the least bit interested in what was happening in the apartment. Besides all these serious problems, which she had to deal with herself at great trouble and expense, she also discovered that the neighbors living below her, as well as those living in the detached houses next door to her building and opposite it, were sworn dog-haters who constantly harassed her, and blamed every bit of dirt or mess in the yard on her dog. But the worst of all was that she felt that her good luck had turned sour. She had never suffered from weakness or depression, and yet a short while after moving into the apartment, she began to feel a distressing weakness, a lack of energy, and inexplicable melancholy. Her financial situation also deteriorated alarmingly, and after earning good money from giving private lessons, she suddenly found that she had almost no pupils. The situation was awful, and the landlord refused to give her any leeway; on the contrary - he threatened that if she did not pay him the rent on the precise date each month, he would have no compunction in throwing her out. Even though she was not the complaining type, her natural optimism began to disappear. She shared her frustration with a classmate. The friend, who in addition to her medical studies was investigating the medical aspects of the pendulum, offered to visit the apartment and check out its energetic condition.

Even as she reached the exterior of the building, the friend sensed through her developed intuition that the energy in the region of the house was not positive. When she entered the apartment, she quickly took out the pendulum. She wandered through the apartment, checking the energy. The negative movements of the pendulum were so forceful that she recommended that Vivian leave the apartment as soon as possible, and even before she did so, she should change the location of her bedroom, because the most negative energies were in the room she had chosen as her bedroom. These energies caused Vivian to be nervous, tired, and weak.

Vivian took her friend's advice. When she decided to switch apartments, she consulted her friend. There were several possibilities that suited her pocket, and were in walking distance of the university. One of the places was in a neighborhood that was considered really bad, since it had the reputation of being full of criminals and addicts. Of all the places, the pendulum indicated this one as the ideal place for Vivian to live from all points of view. Vivian decided to check the place out, and asked her friend to accompany her - with the pendulum. When she arrived at the apartment, the landlord was curious about what they were doing, and showed great openness regarding the use of the pendulum. They wandered through the apartment, checking the energy in the rooms. After a long time, they found that the apartment was excellent for Vivian, but they would need to purify it.

After Vivian had purified the house with incense, and moved in, she found that she was surrounded by marvelous neighbors. The elderly neighbor living across from her would sit in her garden for hours and watch Vivian's dog, feed him, and spoil him. Her next-door neighbor, who was her landlord, turned out to be a friendly and caring person who helped Vivian with any alterations or repairs that she wanted to do in the apartment. Although the apartment was small and simple, Vivian called it "her palace." She lived there for several years, and the moment she could afford it, she bought it from her landlord for a very good price.

Locating missing objects

Even very organized people sometimes find themselves in the frustrating situation in which the very object that they need right now - the car keys, the shopping list, the child's math notebook or the extension cord - has disappeared as though the earth had swallowed it. They are sure they put them down here, in this exact spot, but there isn't a trace of it. Many people begin blaming everyone in the house, reprimanding the child for not putting things where they belong, or start running madly all over the house and turning everything upside down, usually unsuccessfully. Just to add to the aggravation, sometimes after they have finally given up on finding the object, it suddenly shows up, in the most unlikely place or, irritatingly, right under their nose!

Using the pendulum, you can find the object easily and quickly. The method is very simple. If you have a reasonable ability to visualize, imagine the lost object in front of your eyes. Stand and hold the pendulum, preferably in one of the central parts of the house that lead to the rest of the rooms. Hold the pendulum and ask: "Is the missing object [say the name of the item] present in the bedroom? Is the missing object in the children's room?" and so on. When you have received a positive answer from the pendulum regarding one of the rooms, go into that room, stand in the middle of it, and, facing in one direction, hold the pendulum and ask: "Is the missing object in this direction?" Keep turning in a 360-degree circle, until the pendulum produces a positive movement. Note the spot where you were standing, and move to another corner of the room. Repeat the process. The point at which the two lines intersect is the point where you should look for the object. It is amazing with what ease and speed the object can be found.

Taking care of plants

Professional gardeners, amateur gardeners, and even people who do not consider gardening one of their hobbies, enjoy filling their houses and their gardens with a variety of plants. But everyone knows that without proper care, even the strongest and sturdiest plant will wither away. If you are accustomed to using the pendulum, you can save yourself a lot of time searching through professional gardening literature, as well as a lot of disappointment. As organic bodies, plants are good at letting the pendulum know about their character and needs. If you wish to have a lovely garden, or a house full of greenery and life, start using the pendulum while still at the nursery.

Even if you purchase your seedlings, potted plants, or seeds at a good and reputable nursery, the workers at the nursery are not always aware of the precise conditions of the plant you are intending to buy. One of the problems in buying plants is that they may look healthy and fresh, but after a few days at home, they turn out to be weak or sickly. Sometimes the plant was already weak before you chose it, and sometimes it is the unsuitable environment or improper care that caused the plant's unfortunate condition.

When you arrive at the nursery to select plants for your home or garden, take the pendulum with you. When you have seen a plant you would like to buy, hold the pendulum over the plant and ask: "Is this plant healthy and strong?" (If you are experienced at using the pendulum, you can concentrate on the questions without saying the words out loud.) If the pendulum gyrates in a positive response, the plant is basically healthy. If not, go on to the next plant. If you are blessed with strong visualization abilities, you can progress to the next stage, which can save you a lot of trouble. Hold the pendulum above the selected plant, and visualize the surroundings in which you plan to plant or place the plant. When a picture of the spot comes up in your imagination, ask: "Is this plant suitable for the area in which I want to put it?" You may have to change the intended location of the plant, or to choose a different plant, if you insist on that particular spot. If you buy fertilizer and soil at the nursery, check their compatibility with the plant you have bought, or with your house or garden plants.

After you have purchased the plant you wanted, make sure to position it in a suitable location as regards light and sun, temperature, fertilizer, the exact amount of water suitable for a plant of this type, and so on. Use the pendulum to check out all these criteria. The pendulum can help you find the exact spot in which to place the plant (use the 360-degree arc technique described in the previous section), or stand on the spot in which you plan to place or plant the plant, and ask if this spot is the most suitable for the plant. You can also use the pendulum to check how much water and light the plant needs, and so on.

If you take care of your plants using Bach Flower Remedies, or you water certain plants with water in which a crystal or gemstone has been immersed, use the pendulum to check which remedies or crystals are appropriate for that plant. Because of the amazing ability of plants to connect with the environment, we can easily know, using the pendulum, whether the plant feels good or not. The method is very simple. Hold the pendulum over the plant. If it gyrates in a positive direction, the plant feels good. If it gyrates in a negative motion, use it to discover the reason for the plant's negative feeling. This check is important, because it shows the existence of a problem long before the problem manifests itself outwardly. Similar to human beings, feeling unwell often precedes the appearance of an incipient disease.

The pendulum and animals

Choosing an animal is like choosing a friend for life. The fidelity of dogs and cats, for instance, who live in a loving and nurturing environment and enjoy respect and affection from their owners, is amazing and moving, and sometimes nearly incomprehensible to humankind. Animals that are well taken care of tend to become attached to their owners with bonds of unconditional love. Many of them even develop exceptional telepathic communication with their owners, and are very affected by their emotions and moods.

Although we call a person who is responsible for a given animal its "owner," like the dog's "owner" or the cat's "owner" - we must be fully aware that this animal does not belong to us. From the moment we choose to bring it into our homes, we are responsible for its feeding, healthcare, and well-being, but it is never, ever "ours" - rather one of the creatures of the universe whom we have chosen as a friend and companion. It is never an "object" that we can do with as we please - ignore its feelings or even pass it on from one person to the next.

Therefore, the importance of choosing an animal, either a baby or an adult animal, is of very great significance, both for you and for the animal itself. Many people make a mistake and choose an animal whose character is not appropriate for them, for example, naturally quiet and home-oriented people who choose an active and energetic puppy, or people who buy a cat and only after a certain amount of time discover that they are unable to tolerate its unique character. The unfortunate thing is that insensitive people often throw the animal out of the house or give it to another family, which can initiate a syndrome of passing it from one person to the next, from which the animal emerges hur and upset because of the speed with which it bonds with people, and the trauma of the ensuing abandonment. It is not always easy to discover the true nature of a puppy or kitten or an adult animal from observing them for just a few hours. And it is even harder to know whether there will be a good match between your energies, and whether your living environment and your other family members will be good for the animal.

Therefore, when choosing an animal, don't hesitate to employ the pendulum. When you want to get acquainted with it, stand beside it and hold the pendulum over it. This method is suitable for all animals. Fish are very sensitive to the energies around them and will quickly die in an unsuitable place. In general, fish and birds bought at a pet shop are likely, within a few days, to be discovered to be harboring diseases that could not be discerned by the inexperienced eye.

When you hold the pendulum over animals, declare your objective, ask the opening questions, take a few deep, comfortable breaths, and ask the following questions: "Is this animal appropriate for me?" (It is worth asking this question beforehand at home, if you are deliberating between taking a cat or a dog, for instance). Observe the movement of the pendulum. If it produces a strong, smooth, clear positive movement, it is a sign that the match is excellent. If the movement is negative, it is best not to take a chance unless you felt an extraordinary affection for the animal, a strong instinctive feeling of communication between the two of you, and you are willing to commit yourself to making an exceptional effort in order to establish a good relationship with it. You must know and recognize your limitations. If you are going to choose one puppy/kitten from a litter, ask: "Is this the most suitable kitten/puppy for me?" Now continue, and ask: "Is this animal in good health?" If you receive a positive answer, go on to the next question. Concentrate on your dwelling place and imagine it in your mind. If you live with additional family members or other people, visualize them as well. Now ask: "Is my living environment (and the people living in it) suitable for this animal?" If you receive a positive answer to this question too, then you have found a faithful and loving soulmate. Best of luck to you both!

13

Advanced uses of the pendulum

After prolonged use and practice with the pendulum and acquiring a certain amount of experience in operating it, you can expand the areas of its use to include more unusual applications. These uses are ancient, and acquiring experience and practice in them will lead you to discover even more uses, which also constitute ancient knowledge that is likely to come to you intuitively if you make sure to use the pendulum only for positive purposes in a sincere and genuine manner.

Map dowsing

Map dowsing is one of the uses for which the pendulum has become best known, because of its numerous possibilities. It serves many different functions, from finding lost people and discovering deposits of natural resources, to military uses - locating military bases, enemy submarines - and on higher levels of working with the pendulum, for discovering lines and centers of energy.

Even though we are in fact describing something that is not visible to the eye, and which is not possible to detect through regular means, the technique itself is quite simple, and in most cases works well. This technique has been used to discover treasure, water, mineral deposits, quarries, ideal sites for construction or living, lost pets, objects that went missing in an open space, lost persons, and more.

This technique is performed in several stages. It can be done with many different variations after having acquired some experience, and you can determine the method that best suits you. First of all, select the appropriate map and spread it out on a clean, empty table, or on any other level, flat, stable surface.

First stage
Declare your objective, ask the opening questions, and take several deep, slow, comfortable breaths. Now hold the pendulum over each of the four wind directions on the map - north, south, east, and west, and afterward in the middle. Over each point, ask: "Is what I am looking for in this area/direction?" When the compass gyrates in a movement signifying "yes," this is the direction or the general area in which the thing you are looking for is located.

Second stage
Now you must narrow down the possibilities of its location with the pendulum. The most simple way is to ask the pendulum precise questions. Look at the areas in the direction or region over which the pendulum made a positive movement, and begin by asking if the thing you are looking for is to be found in a given city, a certain neighborhood, street, corner, and so on. This method allows you to locate the object with great accuracy.

You can polish this technique as you wish, and although finding the object or the place might take a long time at the beginning, gradually you will be able to find what you want with amazing rapidity. An example of using the map dowsing technique can be found in this moving story of using the pendulum to find a little girl who went missing.

Gail and Leon

Gail and Leon took their three children to visit their family in Nice, France. The days went by pleasantly, until an event occurred that they would never forget. Their relatives took them to a huge mall. They wandered among the departments that overflowed with gorgeous things, making sure all the while that they didn't lose sight of their children. The two older children, Rachel, aged 14, and Brad, aged 10, loved wandering about, trying on clothes and shopping, but minding their four-year-old sister, Anna, was a real drag. She never stopped nagging that she wanted to go home and play, and Gail had to hold her hand tightly to prvent her from running away and touching every item she saw, to the annoyance of the French salespeople.

When they reached the floor of the clothing departments, Gail, Rachel and Anna went to the women's department, while Leon and their host went with Brad to the men's department. Gail was overwhelmed by the abundance of stylish and relatively inexpensive clothing, and asked Rachel to mind Anna while she, Gail, went to the fitting rooms to try on the garments she had selected.

When she came out of the fitting-room a short while later, she did not see Rachel and Anna. Without feeling upset, she strode confidently over to the teen clothing department, knowing how impatient Rachel was. And indeed, there was Rachel standing next to a revolving rack full of shirts, completely engrossed in choosing one. But Anna was not with her. Rachel swore that she had been standing right there beside her only a moment ago. Gail, who was beginning to get a bit frightened, began looking for Anna all over the department, but with no luck. The men soon joined the search. They went over almost the entire clothing floor, which took a very long time, but found no trace of Anna. Gail began to lose her cool. More than once she had heard about children who went missing in large malls and were never found again, and about gangs that hung out in malls, whose objective was to kidnap small children and sell them for adoption. The most terrifying scenes passed before her eyes.

They quickly went over to the information booth. The French clerks understood how they felt, and allowed Gail to page her daughter in English. The clerks broadcast a message that whoever found the little girl should bring her to the information booth, and asked Gail to wait with them while Leon and the host continued the search.

Gail waited for over four terrible hours for someone to find Anna. She lost her cool and burst into tears, demanding that the information clerks keep making the announcement about the lost child over the PA system. When Leon and his French cousin returned to the information booth empty-handed, Gail thought she would collapse. The doctor who on duty at the mall was obliged to give her tranquillizers, and made her sit and calm down.

Suddenly a woman arrived at the information booth, announcing that she owned the New Age store on the first floor of the mall. She requested their permission to try and discover Anna's whereabouts, using the pendulum. Gail and Leon, who felt that they'd exhausted every other possibility, thanked her warmly, praying that she would find their daughter. The woman asked the manager of the mall, who had come to give Gail and Leon moral support, to give her a detailed map of the mall. He immediately went up to his office and returned with the map, which included all the floors, all the levels, and all the stores.

The woman asked the information clerks to clear the

large table that stood in the information booth, and spread the huge map out on the table. She pulled a small cloth container out of her bag, and withdrew a small pendulum made of quartz. She sat down at the table, closed her eyes, and concentrated. After a few seconds, she grasped the string of the pendulum and suspended it over the part of the map of the first floor. She asked: "Is the little girl Anna on this floor?" The pendulum gyrated in a negative movement. The woman continued checking the rest of the floors. Finally, above the third floor of the mall, the pendulum produced a positive movement. Now she knew what floor Anna was on. She held the pendulum over one corner of the map and asked: "Is Anna in this direction?" She continued asking until she found that Anna was on the west side of the third floor. She began searching for the girl by department. When she discovered that the girl was in the cosmetics section on the third floor, she told Leon that he could go down and start looking for the child, while she discovered the exact store Anna was near. Leon and his cousin immediately went down to the third floor and began searching the cosmetics section. As they searched, they heard the information clerk announce that the child should be found near a certain store in that department. The salespeople in the department joined Leon and his cousin, quickly leading them to the shop where Anna was supposed to be found.

When they arrived, they began looking in the area of the shop and inside it. But to no avail. Suddenly, they heard a shout from one of the salespeople who had joined them in the search. At the side of the shop was a large decorative bench made of stone and surrounded by greenery. Little Anna was fast asleep at the foot of the bench, partially concealed by the leaves of a large fern. The overjoyed Leon gently woke her, took her in his arms, and hurried back to join Gail, who went mad with joy. They warmly thanked the woman, who would not accept a financial reward from them, saying that the joy of finding the child was more than enough for her.

Finding out details about a person using a picture

This special technique has been used by pendulum experts for many years, and produces far-reaching results. Using a picture, an experienced pendulum expert can find out many details about the person - his age, date of birth, profession, beliefs, approach to life, social and economic situation, status, and so on. The secret of this technique lies in the fact that a picture serves as a vessel for tuning in to the person portrayed in it. On the face of it, energy has no restriction of distance or time, and it is possible to connect with the energies of a given person from any distance, as is done, for instance, in telepathic transmission between two people. When a pendulum expert concentrates on the picture, he is in essence tuning in to the person's energies, which because of his observation become stronger around the picture itself. The process of tuning in to a certain person also occurs when a pendulum expert sees that person's picture in his mind's eye, but many experts claim that the results of a reading using an actual picture of the person are better, quicker, and more accurate. There are many different techniques for doing a reading using a picture. Below is one of the more familiar and practical techniques:

Sit in a room free of interfering energies (such as the strong energy of a color television, and the like). Place a picture of the person about whom you are doing the reading on a table. Some experts like to write the person's name on the top of the picture, sometimes along with his mother's name. Declare your objective, ask the opening questions, concentrate, and take several deep, calm, slow breaths. Now look at the picture of the person. The questions you must ask are specific to that person, and you need to make use of the information present in the picture (that is, if a young person appears in the picture, start off by asking about a suitable range of ages). Hold the pendulum above the person's picture. If you wish to find out the person's age, for instance, start asking about age ranges: "Is the person in the picture between twenty and thirty years old?" and so on, until the pendulum responds by making an affirmative movement. Narrow the range of ages down until you arrive at the person's exact age. Continue asking additional details about the person by means of yes/no questions, such as: "Is this person presently married? Is this person an American citizen?" and so on. If the pendulum responds to a particular question with a jumpy or static movement, rephrase the question. Sometimes when working with the pendulum in this manner, intuitive messages about the person are received. Check these with the pendulum as well.

Conveying telepathic messages using the pendulum

Telepathy is a form of non-verbal transmission that expresses the ability to transmit thoughts, emotions, and feelings between people. Its existence has been confirmed via numerous incidents, some of them documented. Most people have experienced a telepathic communication between themselves and relatives, good friends, or even animals. You have probably also experienced a strong urge to call a friend whom you hadn't spoken to in ages, just to be told by him that at that very moment he'd been thinking about calling you. These situations are not rare. There are also people who are capable of transmitting complete pictures to other people via their thoughts. One of the methods of facilitating the transmission of telepathic messages is by using the pendulum. In this technique, you need to place a picture of the person to whom you would like to send the message in front of you, concentrate on the picture and on the message you want to send, with your head free of other thoughts, and hold the pendulum above the picture. The movement of the pendulum should indicate to you whether the message was received, since the stronger your message is, the broader and more rapid the movement of the pendulum will be. There are those who apply this technique by using the picture without the pendulum, but many experiments have shown that using the pendulum strengthens the telepathic transmission and helps it to succeed.

Odelia

Odelia, a young woman whose brother is interested in parapsychology and uses a pendulum, told the following story from her own experience in transferring telepathic messages using a pendulum:

Odelia and her girlfriends went to a cultural festival in another state. They planned to stay there for four days, so that they would have time to see most of the performances, and would sleep in sleeping-bags. Odelia's mother was to arrive on the fourth day, the last day of the festival, and spend some time with them.

The first day was a lot of fun, intensive and full of surprises. When they decided that it really was time to call it a

Pendulums

day, they fell into a deep sleep immediately, exhausted from all their activities. At that same time, at 04:00 am, Odelia's mother, who was at the beginning of her ninth month of pregnancy, started having labor pains, and was taken to the hospital to give birth. Trevor, Odelia's brother, who was 24 at the time, knew that Odelia would be very disappointed if she found out that she had missed being with her mother during the delivery, and welcoming the new baby. He had no way of getting in contact with Odelia, because she and her friends had not given any exact details about where they would be. He understood that he would have to wait until Odelia called home, but, because he knew his sister, he knew she could happily go through the entire four days without calling home even once. The only way open to him to contact Odelia would be telepathically. He decided to utilize the pendulum. While his parents went off to the hospital, Trevor insisted on staying at home and waiting for a call from Odelia. He took a recent photo of Odelia, held the pendulum over it, and began asking Odelia in his thoughts to phone home. To his surprise, the pendulum began to gyrate rapidly over the picture, making a smooth broad clockwise movement. He felt that the message had been transmitted successfully.

Odelia awoke at 06:00 am feeling an inexplicable need to phone home. While her friends were still sleeping soundly, she began to look for a callbox. Even though she knew that her parents would still be sleeping at that hour, her need was so great that she didn't care if she woke them up. To her surprise, the receiver at the other end was lifted immediately and she heard her brother's excited voice telling her about the birth. Of course, they arranged to meet at the hospital. Odelia took the first bus in the direction of the hospital. After the initial excitement of the delivery, she and her brother sat in the hospital cafeteria, and he told her how he had "transmitted" his desire that she should call home. She told him that she had really felt as though something was waking her and making her call home immediately.

14

The dowsing rod

The dowsing rod or divining rod is an ancient tool known in both the East and the West, and its the use passes like a unifying thread through the ages. In fact, the pendulum as we know it today developed from the various dowsing rods in ways that are not precisely known, and many people who use the pendulum use one or more types of dowsing rod as well. There are many different types of dowsing rods, and their use is extremely common for purposes of finding water and underground water reservoirs, oil and mineral deposits, lost objects, missing persons, and buried treasure. On a higher plane, dowsing rods are used to detect the outline of the human aura and to discover blockages and holes in it, and to find lines of energy and energy centers.

Already in the Bible we can see how dowsing rods were imbued with divine powers. The Egyptian Pharaohs worked miracles with their rods, Moses' rod turned into a serpent in Aharon's hand, Aharon's rod bloomed, and above them all, there is the the famous story in which Moses struck the rock with his rod. This tie between water, a gift from the universe without which man cannot live, and the dowsing rod, has continued from ancient days until the present. Because of this important use of the dowsing rod, its users were fortunate, and they succeeded in using it even during the perilous days of the Dark Ages. During those days of witch-hunts, when any hint of "sorcery" was punishable by death at the stake or by terrible torture, the Church turned a blind eye to those dealing in rhabdomancy, or "dowsing," as divining using a dowsing rod is called today, and allowed them to carry on their activities.

An interesting bond forms between the dowser and his rod, and many dowsers are likely to hold onto one rod throughout their entire lives, because they feel that this rod has learned to understand them. An interesting point that has been discovered by many dowsers is that when there is a genuine need to operate the dowsing rod - for instance, in a case when there is a water shortage and an underground well must be found - the rods tend to work in an extremely smooth, rapid, and accurate manner. On the other hand, when they are used for showing off, they are likely to disappoint. This situation is not uncommon with the pendulum, either. In fact, it occurs with the use of any spiritual thing. The thing is designated for a particular purpose, and not for bolstering our ego. Therefore, if you have practiced using one of these search instruments - the pendulum or one of the kinds of divining rods - and you are not able to operate a particular type of rod, don't worry. Continue practicing and working with the search tools that work for you, and know that the moment you need to operate this instrument for an important purpose, it will surprise you and work very well!

Types of divining rods

There are many types of divining rods with different shapes and purposes, but all told, they all serve the same objectives very well. It is known that a certain shape of rod is more suitable for a particular person, while a different shape is likely to be less appropriate for him, and he will have trouble operating it. It is therefore worth trying out several types of rods in order to find out which one works best, most comfortably, and most smoothly for you.

The "L" Rod

The "L" rod is an L-shaped rod that is held by the short branch of the L. They are usually employed in pairs (two rods) - one in each hand. To start out with, they can even be made from copper wire or thick straight wire that has been bent into an L.

The "L" rod should be held like a gun ready to be drawn, ensuring that the fingertips do not touch its long branch. (Remember, the rod is held by its short branch). Throughout the entire search, the operator must remain concentrated on the objective and maintain purity of thought, that is, not allow opinions and thoughts to influence the search. If you activate your own thoughts while you are looking for an underground water pipe, reservoir of water, a given mineral, or anything else, convinced that it should be in a particular place, your thoughts will affect the movement of the rods, and the result will be inaccurate. Remember that you are interested only in the truth! There are operators who habitually murmer mantras such as: "I am looking for the water pipe, water pipe, water pipe, water pipe," in order to remain focused on the objective and not to allow thoughts and opinions interfere with the search and affect it.

When the operator stands above the spot where the object of his search is present, the two rods intersect (that is, they veer toward each other), or they veer outward. In any case, the movement will occur spontaneously, without the operator activating or swivelling his wrists.

To begin with, you can look for the pipelines that convey water to and from your home, if you know where they begin. When you are standing over the point where the pipe exits your house, hold the "L" rods in your hands as though you were holding a gun, and declare: "I am looking for the pipe that carries water to my house, and I ask that the 'L' rods cross when I reach it." If you do not succeed the first time, don't give up. Sometimes some practice is required to operate these tools. But remember, it is very important to clear yourself of any thoughts which are likely to affect your work with the rods.

The "Y" rod, or the forked rod

This rod is the most famous form of the dowsing rod, and its use is the most common. It is shaped like the letter Y, that is, two branches forking off one stem. The arms of the rod must be equal in length, about the thickness of a pencil in diameter, and about 30-45 cm. long.

The "Y" rod has the greatest success when the dowser is looking for a certain point on the ground. The "Y" rod is usually prepared from wood: some people believe that the best trees for making it are apple trees and willows. On the other hand, there are dowsers who use any kind of tree, or even a "Y" rod made of plastic. In principle, any hard material should be fine. There are operators who find their "Y" rods in the forest, waiting and ready for them, and there are those who find them in the street, a Y-shaped object made of some kind of plastic or other substance. For some reason, the rods that are "found" are considered to be luckier and more faithful to their operators. Choose whatever type of material you fancy, as long as the shape is right. When cutting a "Y" rod, you should ask the tree's permission and thank it. Nature is happy to help us for any positive purpose, but we must not forget her generosity or take it for granted. If you use a pendulum, it is worth using it to check whether this is the rod that is best for you (before you cut it off the tree!).

After you have found the "Y" rod that is appropriate for you, hold it by its two arms with your fists closed around the arms of the rod, and pointing upward (that is, with the back of your hands facing upward), and your thumbs facing outward. The rod's single arm should face outward and upward. Look for the suitable position in which you feel that the rod is completely balanced. (There is such a balance point, and it is easy to find.) This is the "Y" rod's search position.

The bobber rod

The bobber rod is a rod shaped like a backward fishing rod - the thinner part of it is used for grasping, and the thicker part moves freely. This instrument is the least common of the four tools used in searching for and detecting things: the pendulum, the "Y" rod, the "L" rod, and the bobber rod.

You can make yourself a rod like this out of a simple, flexible fishing pole, or a long flexible rod whose one end is thick and which tapers off and becomes thinner toward the end.

The bobber rod is held at its thinner end, in either one or both hands. Its search position is immobile, its "yes" movement is usually an up-and-down movement, while side-to-side movement indicates a negative answer. However, to be sure about the movements of your specific bobber rod, do the following:

When you find the branch or fishing pole that seems suitable for you, hold the pendulum and declare your intention to the pendulum: "I want to know if this rod is appropriate to serve as a bobber rod." Ask if you may ask this question, if you are allowed to, and if you are ready.

After you have received positive answers, hold the pendulum over the rod and ask: "Is this rod appropriate for me to use as a bobber rod?" If the answer is affirmative, you can use the rod. It is important to remember that, if you plan to cut or tear a branch off a tree, the check should be performed before doing so. (Disrespecting nature and Mother Earth will produce bad results when you later request that Mother Earth give you water, oil, minerals, or anything else that you search for using the rod! One doesn't soil one's own nest!) After cutting off the branch, the tree must be thanked for its gift. (The energy with which the rod was taken will affect the rod itself, therefore it must be pure and infused with gratitude.)

After you have found the appropriate rod, you will need to find the correct position to hold it in. Hold the rod in one hand, with its thin end placed between the thumb and the forefinger, while the rest of the fingers are placed a little bit inwards; or, if it's more comfortable, grasp it in both hands, with the bobber rod's thinner end placed between the two lightly fisted (closed) hands, with the fingers of the right hand tucked in under the fingers of the left hand, and the end of the rod held under the right hand. There are those who cross their thumbs over their fisted hands, but care must be taken that the thumbs do not touch the rod itself. Left-handed people sometimes place their left hand under their right hand, but you will have to find the position that most suits you.

After you have grasped the rod correctly and its thicker end is pointing horizontally and rising slightly upward, it is time to tune it. Hold the rod and request: "Show me your 'yes' movement." This movement will probably be a vertical movement, up and down, or a sort of light skipping up and down, or a movement from side to side. After you have found the "yes" movement, ask the rod to show you its "no" movement. (This movement is usually from left to right or right to left, similar to shaking one's head "no," but in the end the rod will show you its movements.) In any case, one movement will be up and down or down and up, and the opposite one will be from right to left or left to right.

The bobber rod is also used to find out to what depth you will have to dig in order to find whatever it is you're looking for (water, oil, minerals, quarries, lost objects, and so on). In order to determine the depth, hold the rod above the place where you discovered that the thing you are looking for is located. Tell the rod that each "yes" movement is to signify three, five, or ten meters - whatever is convenient for you. When the rod begins to move, count the number of movements it makes, from the first one until it stops moving and returns to its search position or its "no" movement. (The "no" movement is not to be counted!) Multiply the number of movements by the number of meters you had selected (for example, if you requested that each movement signify ten meters, and it moved five times, the thing you are looking for is at a depth of 50 meters below the surface). If you want to know the exact depth of its location in individual metres, change the search code and tell the rod that each "yes" movement is to signify one meter (or half a meter, whatever is convenient). Count the number of "yes" movements until it goes back to its search position or moves in a "no" movement. Multiply this by a meter or half a meter, according to whichever code you gave it, and thus you will know the exact depth at which the object of your search lies. You can even search to the point of individual centimeters using this method. All you have to do is retune the rod each time to the length code you have chosen, and note the number of times the rod makes its "yes" movement.

The work of a dowser using one or more types of divining rod or pendulum is done with different techniques, combining different levels of work. At each level of work, the dowser has different work techniques at his disposal and specific abilities. When you start working with these instruments, you can check which level you are at and gradually rise from one level to the next, beginning with the most down-to-earth level of work, in which it is possible to locate something only if the dowser is standing directly above it, to the most highly spiritual level, in which the divining rod serves to locate tears and perforations in the aura and energy lines on the earth's surface. It is possible that you will find yourself naturally starting at a higher level, but there are also dowsers who are not aware enough to rise to the higher levels of use. Upon increasing the level of awareness, new techniques are discovered and new abilities manifest themselves.

First technique: Finding objects and things when the dowser is standing over them

In this technique, the operator using the pendulum or dowsing rod can find objects, water, or minerals, and so on, only when he is standing directly over the object of his search. Here, the search tools act like radar, and they pick up the frequency emitted by the object, the water, or the deposit. This technique is very simple to perform and does not require a great deal of experience or knowledge. It is primarily used for checking and acquiring confidence that this is in fact the place where one needs to dig, so as to prevent futile efforts.

Second technique: Finding objects and things when their location is not known

In this technique, which requires somewhat more skill, experience, and awareness, the dowser can begin searching even if he has no idea of the location of the thing he is looking for. He can stand at the edge of the area in which he is carrying out the search, holding the pendulum or the dowsing rod in his hand, and ask to know in which direction the thing he is looking for is, at a given depth that he determines. The question can be asked in this manner: "What is the correct direction for finding an oil deposit in this region, and at a depth of less than ... meters underground?" When using the pendulum, the operator begins turning slowly in a complete circle, and when he is standing in the correct direction, the pendulum begins moving in a "yes" movement, or leaves its search position. When using the "Y" rod, it should be held in the search position while you revolve slowly in a circle. When you are standing facing the right direction, the "Y" rod will leave its search position and move upward or downward. When using the "L" rod, the operator holds one rod in the search position (like a drawn gun) and turns in one direction. The "L" rod will move in the direction of the item you are looking for, and will stick to this direction, as though it were glued to it, even if you continue turning. This technique works when you are not standing over the thing itself, but at a visual or walking distance from it.

Third technique: Map dowsing

This technique is employed when the object of the search is not within your field of vision or within walking distance from you. In this technique, you will need to use a map to serve as the object of the search, that is, your area of search is marked on it, and it needs to be as accurate as possible. The general search can be started on a less detailed map and gradually, when you have reference points such as the city in which the thing you are looking for on the map of the country is, you will need to transfer to a city map, and from there to find the street, and so on. This technique is very similar to map dowsing using the pendulum, and the uses made of it are extremely numerous. As you practice this technique more and more, you will find that it gradually become easier and more natural, and leads to wonderful and far-reaching results. This technique is used by most of the great dowsing experts, as well as those working for large companies, governments, or military organizations.

Fourth technique: Technique for connecting with the search instrument

This technique requires a certain degree of belief in your search instrument and good communication with it, and is considered one of the superior techniques for using these tools. In principle, it constitutes the true objective for which these instruments were created - to raise the operator's intuitive level to such a high degree that he no longer needs the search tool! This technique was attributed to the people of the ancient secret and spiritually advanced cultures of Atlantis and Lemuria. After you have used one of the search instruments for a long time - the pendulum, the "Y" rod, the "L" rods, or the bobber rod - you will be able to perform a search without the tool itself! In fact, the purpose of these tools is to serve as mediators between human beings and cosmic knowledge. An explanation of this can be found in the chapter discussing the manner in which these instruments work (in the chapter, "How do the pendulum and the divining rod work?"). This technique is performed in the following stages:

Checking capability: Sit in a place free of any sort of interfering energies, check that your arms or legs are not crossed (sit in an "open"' position), relax and take deep, slow, comfortable breaths. Now ask: "Can I operate without tools?" Relax, take several deep breaths, and close your eyes. Imagine a screen or an open window on your forehead between your eyebrows (the location of the third eye). Look into this place, and see your pendulum or dowsing rod in its search position (it is easiest to visualize the pendulum in this technique). Look at the pendulum and see it move from its search position into your "yes" movement, and return again to the search position. Continue asking: "Am I allowed to do this? Am I ready to do this?" If you received an affirmative answer to your questions, you can attempt searches or readings without instruments. All you have to do is to visualize your search tools in this manner, declare your objective, ask the opening questions, and ask the question that you wanted to ask. You will see the correct answer in your mind's eye.

It could be that the visual way is not suitable for you, because one of your other senses happens to be more developed. If you can "hear" the answer, or "feel" it (some people use their finger as a pendulum, and sense an inner movement of "yes" or "no" in the finger itself), use whatever sense is appropriate for you in order to do a search or a reading without instruments. This is a very high level of use of these tools, but after prolonged practice with the instruments and good attunement to them, it is very easy to achieve. In fact, we are meant to and are capable of arriving at the pendulum's and search tools' answers without the instruments themselves because, as was stated, they are only mediators between us and our Higher Selves, or Cosmic Knowledge. If you did not succeed, it doesn't matter. Continue using the instruments you have, and from time to time, try this technique again. At some stage, you will probably succeed.

Sources

I wish to mention two books from which I learned the uses of the pendulum:

1. *Pendulum Power* - G. Nielsen & J. Polansky (Aquarian) - a fundamental book that exerted a great influence on my pendulum studies. The book contains a detailed bibliography.

2. *Spiritual Dowsing* - Sig Lonegren (Gothic Image) - a fundamental book in the field of the rod. The book contains a detailed bibliography.